MOVEMENT IN LONDON

TRANSPORT RESEARCH STUDIES AND THEIR CONTEXT

GREATER LONDON COUNCIL

Published by the
© Greater London Council
The County Hall, London, SE1
1969

Publication 7168 **0251** 1

Price **£15** (postage extra)

Design and production by
GLC Supplies Department,
Printing and Graphic Design Division.

Printed at The Curwen Press Limited.

Bound by C. & H. T. Evans Limited.

Maps and diagrams prepared
by Woodcote Press Limited.

Blocks supplied by
Art Reproductions Limited
and W. F. Sedgwick Limited.

Maps and diagrams reproduced from
or based upon the
Ordnance Survey map with the
sanction of the Controller
of HM Stationery Office.

Acknowledgements

This report describes work that has been completed over a period of six years. During this period the London Traffic Survey and the London Transportation Study were carried out by Freeman, Fox, Wilbur Smith and Associates as consultants to the Greater London Council and its predecessor, and to the Ministry of Transport. During this period also, local government organisation in London underwent major changes. The professional expertise of the consultants ensured that there was no loss of impetus during the inevitable adjustment in the administrative control of the project. The consultants developed many new techniques that are now being used by the Council's staff and will thus benefit planning in London for many years to come.

While the major credit is due to Freeman, Fox, Wilbur Smith and Associates, they would be the first to point out that their work had benefited from the active support and cooperation of many other individuals and bodies. In particular a major contribution has been made by officers of the British Railways Board and Regions and of London Transport Board.

During the London Transportation Study substantial contributions were made by officers of various departments of the Greater London Council including particularly the Valuation and Estates Department, the Planning Department and the Department of Highways and Transportation, as well as by officials of the Ministry of Transport. Research Projects Ltd. were consultants for smaller but nevertheless significant parts of the work concerning modal choice.

The work described in this report subsequent to the London Transportation Study has been undertaken by officers of the Research Group of the Department of Highways and Transportation in conjunction with colleagues in other departments of the Council.

The firm basis that has been established for the further development of transport policy in London owes much to the personal contributions of many individuals both inside and outside the organisations mentioned above.

Movement in London

PART 1

Chapter 1 Policy and research

PART 2

Chapter 2 Basic relationships in 1962 (Phase I)

Chapter 2 **Basic relationships in 1962 (Phase I)** *continued*

Chapter 3 Potential demand in 1981 (Phase II)

Chapter 4 Constrained demand in 1981 (Phase III)

Chapter 5 Work undertaken by the Greater London Council

Appendices

Appendix F Economic evaluation

Appendix G Trip pattern in 1981

Glossary of technical terms

Movement in London

This publication describes the transportation planning work started by the London County Council and continued by the Greater London Council in conjunction with the Ministry of Transport. The results of earlier phases have been published as Volumes I and II of the London Traffic Survey. The purpose is to relate the earlier work to Phase III (the London Transportation Study), to present the results of Phase III itself, and to present the results of studies completed by the Council during and since Phase III.

The report is divided into two parts. Part 1 describes the policy context in which the research was started, illustrates how the research has affected policy and shows how the development of policy has extended the field of research. Part 2 is a technical presentation of the results of the research in the order in which the studies were made. It is based on detailed reports prepared by the consultants and the GLC and only repeats the findings of Phases I and II in so far as this is necessary for an understanding of the later work of the consultants and the Council.

The Council has published the Greater London Development Plan. This report provides additional background information which supports the conclusions reached in the Studies Volume of the Development Plan and in Section 5 of the Written Statement.

Part 1

1

Chapter 1 Policy and research

The background

1.1 Between 1955 and 1960 the LCC increased its annual expenditure on road improvements from less than half a million pounds to £4½ million. This nine-fold increase was urgently needed to meet the growing traffic demands; but it was already clear that the rising standards of living which produced the need for an increased road programme was also changing the nature and characteristics of movement in London. A comprehensive basis was required for future road planning.

1.2 In recognition of this need a study was proposed by the LCC in 1960 with the following objectives:

a. To survey origins and destinations of traffic movements within an area to be defined, but not much larger than the LCC area.

b. To survey journey times.

c. To develop a method of estimating the amount of traffic on a road network.

d. To indicate areas where quality of travel was poor and capacity limited.

e. To present the results in a form suitable for the preparation of a comprehensive road plan for the County of London.

1.3 Development of the proposals with the Ministry of Transport and consulting engineers extended the geographical boundaries of the study to include the built up area of London and included studies of commercial vehicle movements and public transport within the scope of the work. The fact finding was also extended so as to provide more of the information that would be required to forecast future patterns of movement. It is significant that even before conclusions had been reached about the reorganisation of local government in London the need to study transport comprehensively led to a study encompassing the main built up area. The traffic survey area was over 940 square miles and the GLC area is 616 square miles and lies wholly within the survey area.

1.4 In 1960 the policy of the LCC was directed towards the difficulties being increasingly encountered on the roads. With petrol rationing ending in the early 1950's, and increased expenditure on private vehicles, there had been a 50 per cent increase of traffic in ten years and speeds in central London during 1958 were the lowest ever recorded. All the available evidence in this country, and experience from developed nations abroad, indicated that a travel explosion was closely associated with rising incomes. Dramatic increases in car ownership, car usage and the demand for road space were associated with increasing affluence and lower density living. What was missing in London was any detailed knowledge of the extent of this demand, the implications of meeting it or the consequences of not meeting it.

1.5 The London Traffic Survey was approved as a first and vital step to enlarge understanding of the complexities of movement in a great city. It concentrated on the major issues facing decision makers then and today: how to meet the challenge of the motor vehicle in an established and valued city that has historically developed in conjunction with a comprehensive public transport system.

Establishing present movement patterns

1.6 The results of the survey have been published previously and are summarised in Part 2 of this report. The findings had two principal uses. First they provided a basis for greatly improving the understanding of layman and expert alike about movement in London. Secondly, the data provided a firm basis for defining numerically the relationships between such factors as population, employment, shopping and economic activity and the use of the road and public transport systems. Some of the simplest findings were the most significant and yet could only be firmly established once the comprehensive information had been accumulated.

1.7 London was shown to be an urban region in which the everyday movement was accounted for almost entirely by the resident population. In 1962 ninety per cent of all the 15·8 million person journeys per day by private and public transport started and ended within the London of the survey area. Only 6·5 per cent of the travel was accounted for by non-residents and most of those had business in London. One journey in one thousand was truly a 'through trip' that had neither origin or destination within the survey area. Immediately it could be concluded that London's traffic problems were not to be solved by diverting movement around the capital. A road system designed to serve the needs of Londoners was the principal requirement but this also had to be consistent with the developing pattern of roads in the South-East of England.

1.8 The survey clearly demonstrated the roles of the different transport systems, which have not changed significantly since. Central London, covering ten square miles (1/60 of the GLC area) with 30 per cent of the total employment, depends for its accessibility on an efficient public transport system. During the morning peak period (8 am–10 am) 35 per cent of the workers enter central London by British Rail services, 35 per cent by underground services and about 20 per cent by bus, and many workers use more than one type of public transport. In particular about one quarter of the passengers using British Rail services continue their journey by underground. Only about 10 per cent of the journeys to work in the centre are by car. Private car journeys are however a much greater proportion of the total peak hour traffic on the road system leading to and within central London.

1.9 Outside central London, employment is widely spread with about 20 per cent of all London jobs within the remainder of the former LCC area (107 square miles) and the other 50 per cent spread over the rest of the GLC area (500 square miles). This contrast in density of development between the core and the inner and outer suburbs of London is reflected in the use made of transport. In London as a whole 90 per cent of the trips made are by bus and car and therefore depend on the road system. Almost all freight movements in London are by road and these amount to more than $3\frac{1}{2}$ million trips per day. About 2 million of these are concerned with retail deliveries ranging from bread to television sets.

1.10 The rail services are predominantly used for carrying workers to central London and their demand is therefore concentrated into peaks. The survey showed that over 70 per cent of the British Rail passengers travelled in the four hours 8 am–10 am and 5 pm–7 pm, 60 per cent of the Underground trips occurred during the same hours, 45 per cent of the bus trips and 35 per cent of the car trips. Clearly the higher the peaks the more inefficient the use made of men and equipment over the day as a whole. This is a cost that follows from the concentration of economic activity.

The factors behind travel

1.11 Information about the current transport situation is important; but the most valuable contribution made by the survey was the opportunity it provided to explain travel in terms of the economic and social life of the capital. To know how many people use buses and how many cars are on the road is one thing. To be able to relate the number of people in buses and cars to the characteristics of population, the density of development, the types of employment and the prosperity of the city provided vital knowledge not previously available. It meant that future travel could be estimated in terms of those features of life that can be influenced by planning decisions and not solely on the basis of past trends in transport itself.

1.12 Almost 90 per cent of the trips made in London start or end at home. The social and economic characteristics of households therefore have a significant bearing on the number and type of trips made. Two factors were shown to have a predominant influence, namely, household income and car ownership. As more people earn more money, they spend a greater proportion of their budget on transport whether they own cars or not. However, as affluence spreads so does car ownership. The car is not solely, or even predominantly, a status symbol but a means of increasing family mobility in a city providing more and more variety and opportunity for work and leisure. Increasing car ownership means more than direct competition with public transport. It means potentially more travel because, given the freedom to do so, car-owning families were shown to make twice as many journeys as families without cars even when both are as well off financially.

1.13 Detailed analysis shows that the extra journeys people make because they have cars are largely associated with trips for business, shopping and social purposes. Some of these trips take place outside present work hours, but many contribute to the pressure on the roads at busy times.

An initial look at the future

1.14 The survey findings, published in 1964, showed the complex relationships between population, employment, household incomes and movement in London. In 1962 less than 40 per cent of homes owned a car so the potential increase in car ownership was large. The possible effects of this were not obvious and the relative rates of increase in demands for movement, particularly by road, and the desirable rates of investment in new facilities had still to be clearly established. London could not be conceived of without an extensive public transport system but it was clear that the future demands on it must be different from the past. How different and in what ways could only be established by making forecasts of future travel demands given certain planning assumptions. If all the interactions between population, employment and travel were to be taken into account an extensive analysis was required. Phase II of the LTS was designed to make this analysis and provide initial travel forecasts.

1.15 London is an old and well-established city region and it has for many years been a generally accepted aim of policy that the population of the built up area which is now Greater London should not increase beyond 8·0 million. This was the planning assumption made in the initial and subsequent travel forecasts and was consistent with the statutory development plans existing at the time. A population of 8·0 million in the GLC area by 1981 represented an increase of 2·5 per cent over 1961.

1.16 Employment in the GLC area was expected to be 5·0 million by 1981, an increase of 11 per cent over 1961. The other important planning forecast used as a basis for the travel predictions was future household income. An annual growth rate of income, after allowing for inflation, of 2·8 per cent was assumed. This corresponded with government forecasts and was the mean rate achieved over the previous 20 years despite the vagaries of economic fortune.

1.17 The significance of the planning forecasts is not their absolute values, although these are important, but their relative growth; population 2·5 per cent, employment 11 per cent, but household income 70 per cent. Thus the future of London on these assumptions was conditioned by large changes in purchasing power within a context of small changes in population and employment patterns and a physical environment that inevitably could change only slowly. The effect of more recent trends in population and employment are discussed later but it is important to appreciate that economic factors have by far the more significant effect on travel.

1.18 The travel forecasts made in Phase II of the LTS indicated a potential doubling of the number of cars so that 70 per cent of the households could have access to at least one car by 1981. The combined effects of increased income and car ownership were shown to be a 50 per cent increase in total travel including car, bus and train. Most of this increased movement was expected to be associated with twice as many shopping and social trips and a 50 per cent increase in business trips. Travel to work in these forecasts represented 35 per cent of total travel compared with 50 per cent at present. Much of the increased travel was expected to take place outside present peak hours and because the trips were associated with activities more often located in the suburbs the potential increase in the demand for movement outside central London was forecast to be 60 per cent compared with a 20 per cent increase to and from the centre.

1.19 As household income increases the demand for consumer goods and services also grows rapidly. There is a corresponding increase in the tonnage of goods to be moved on the roads between manufacturers, wholesalers and shops. How many additional lorries are required to carry the goods will depend on changes in the size of lorries and developments in packaging and freight handling techniques. The forecasts made in 1964 for Phase II indicated that up to 75 per cent more goods vehicle trips might be made on the roads of London by 1981. There are now signs that increases in the size of vehicles may allow the growth of goods movement to be handled with fewer vehicle movements than originally predicted. But the importance of keeping London competitive by avoiding the growth in costs entailed by delaying goods movement still remains.

1.20 A dominant feature of the forecasts is the large increase in travel demand expected outside central London. The patterns emerging are more than a reflection in transport terms of increased activity, they are also a symptom of profound social change, the coming of an age in which the overwhelming majority of citizens will have access to motor cars and in which increasing quantities of goods are distributed by road. The problem of accommodating even a substantial part of this total demand for movement is formidable. To serve movement to and from central London advantage can be taken of both the extensive rail system and of the road system. But movement outside the small central area is distributed over the whole region and has nothing like the same concentration as that to and from the centre. This movement can only be served by bus and private vehicles which depend for their standard of performance on the quality of the road system.

1.21 The existing road system has a predominantly radial pattern and does not match the changing pattern of movement in London that the research studies highlighted. The existing road system cannot accommodate any substantial increase in demand. Even small increases can often only be achieved at the expense of using roads with a residential element to their character. The extent to which there is a conflict between road capacity and demand depends both on the degree to which demand is restrained and on the quality of movement desired for that demand which is in fact served. The more existing capacity is used the greater the degree of congestion and the lower the quality of service offered by the road system. This decline in quality affects different journeys in different ways. Short journeys may be affected less than the longer (and often more important) strategic journeys. Irrespective of whether the total future demand for road space is met in full, if any major increase in the *quantity* of travel by road is desired and if at the same time the *quality* of road travel is to be improved by the segregation of long and short journeys, a system of limited access roads is required.

1.22 Phase II of the LTS research therefore concluded by examining a system of urban motorways consisting of a ring and radial pattern. The capacity of the system was not defined but rather corridors for movement were drawn with assumed travel speeds and access points to the existing road network.

1.23 The pattern of motorways evaluated was not revolutionary. The pattern was constrained by existing facilities, such as the North Circular Road, which are assets too valuable not to be integrated into a future system. It was constrained by an already declared policy of central government to build a national system of motorways, many of which penetrated the London area in a radial direction. It was constrained by the available opportunities to cross the river and the necessity to avoid areas of open space such as Regent's Park and Hampstead Heath. In concept the pattern was similar to the rings proposed by Abercrombie and Forshaw in the County of London Plan of 1943, with the omission of the innermost ring. To evaluate a concept that has developed over many years with the latest and most comprehensive information and techniques is sound common sense. To propose building Abercrombie's rings without evaluation because it was an old idea would be wrong. Equally, to disregard them because it was not a new idea would be wrong, particularly since the pattern corresponds with the need for orbital roads which the survey has highlighted.

Traffic: Environment: Restraint

1.24 The initial forecasts in Phase II were not prejudiced by any preconceived notions of restraint except in so far as it was generally agreed that central London could not be served predominantly by road. The purpose of the forecasts was to estimate demand on the basis of how people would use cars in the future if there were no more restraints than existed in 1962. To forecast this demand does not mean it has to be met in full, especially if to do so means the destruction of the very urban fabric the new roads are intended to support.

1.25 Not to meet the demand in full implies either management measures that authorities can apply effectively or a policy of inaction leading to control by congestion. Congestion is not only inefficient but destructive of the total environment. While substantial moderation of demand at critical times and places is required, a balance must be struck between planned restraint, congestion and extra facilities.

1.26 How the balance is achieved must inevitably be conditioned by the alternatives available, the resources available and the enforceability and fairness of the restraint measures. Different areas with differing circumstances are likely to find different degrees of restraint acceptable and possible in terms of resources and environment. In particular it is generally recognised that restraint of car use in the centre of London is inevitable, and acceptable, in view of the physical environment of the centre and the alternative means of access by public transport. But the same conclusion does not necessarily apply to the inner areas of London and the major outer suburban areas. If no investment in roads were made here and all future demand in these areas were to be entirely regulated by congestion the effects for London as a whole would be likely to be much more harmful in impeding commercial vehicles, business and shopping trips and generally by restricting the economic and social advantages of mobility. In 1966 the GLC adopted policies which combined the elements of restraint in central London with the provision of a network of primary roads to meet the increasing demands for movement throughout the region.

1.27 In February 1966 the Council approved a major step forward in management policy involving the control of all kerbside parking in an area of 40 square miles of central and inner London. When fully implemented this will provide a greater measure of control in terms of time and cost, and over a larger area, than any other major city in the world has achieved. This policy not only protects the viability of many business and shopping areas and protects the interests of residents, but in addition provides a means of regulating traffic flow by influencing the use of private cars and public transport particularly during peak periods.

1.28 In July 1966 the Council adopted as a major long-term policy objective the creation in London of a primary road network with a basic orbital pattern. This was consistent with the conclusions of the research studies and took full regard of the future of public transport particularly in serving central London. The capacities to be provided, the priorities to be accorded, the timing and the detailed economic costs and benefits were to be assessed as soon as possible.

The research continues

1.29 The results of Phase I and II of the LTS provided a basis for broad strategic policies on restraint and road investment. These policies required much important detail to be filled in and the research findings themselves gave rise to deeper and more complex questions. In particular it was necessary to know what levels of restraint were associated with different levels of investment, the effect of restraint on travel patterns, the costs and benefits associated with differing degrees of restraint and investment and the corresponding implications for public transport.

1.30 Transport planning research studies in other cities had not previously examined such complex issues, so new technical methods had to be evolved. The development and application of these methods formed Phase III of the LTS. The methods developed, their scope and their restrictions, and the results of their application are described in Part 2. The new methods were used to evaluate three road plans and two public transport plans.

1.31 The three road plans evaluated were selected so that the results obtained would be as significantly different as possible when one plan was compared with another. The lowest investment plan included only an outer motorway ring and certain committed projects, whereas the highest investment plan included additional facilities beyond the objective approved by the GLC in 1966.

1.32 Different considerations applied to the public transport plans. London already has a comprehensive public transport system so that in the context of existing Green Belt policies and small changes in population and employment the public transport problems are mainly associated with improving the quality of service rather than large increases in capacity to meet growing demands. The emphasis of the studies was on assessing the potential market for different forms and quality of public transport service and the significance of different levels of road investment on public transport services. Two public transport plans were developed each providing for improvements in travel speeds particularly by rail, and including the Victoria Line and the proposed Fleet Line. The higher investment plan featured an extension of British Rail service through the central area with the purpose of reducing the necessity for interchange at existing main line terminals and thereby improving the quality of public transport service to the user. Both plans also included the provision of British Rail and London Transport links to Heathrow Airport.

The effects of alternative levels of investment in roads

1.33 A major part of the technical work of Phase III of the LTS was concerned with assessing what Londoners would do in future if their first choice were to use their car but restraint policies, or lack of road capacity, prevented this choice being made. Very few opportunities exist in the world to study such a situation. The methods developed do not, therefore, rely directly on observed data but on the best procedures that could be devised by the consultants in association with their clients. The results of the studies are therefore of most significance in comparing the relative effects of different plans rather than for the justification of one particular plan. Nevertheless they are the best results available and are derived from methods as advanced as any used in comparable studies elsewhere.

1.34 The low investment plan represents an investment of £370 million at 1966 prices. This implies a high degree of restraint on the use of private cars, some 40 per cent of the potential car trips not being served. The number of trips served by road would be an increase of 26 per cent over 1962 levels and public transport would serve 14 per cent more travellers. It is unlikely that the administrative techniques available and their associated man-power demands could sustain a 40 per cent level of restraint. Congestion would almost certainly have to be an important element in controlling demand in a road plan of this size. The congestion levels associated today with central London could become much more widespread. Evidence of this trend is already apparent in the increases of traffic in inner and outer London with associated decreases in average speeds.

1.35 The medium investment plan represents an investment of £1,650 million and implies that about 6 per cent of potential car journeys would not be served and that there would be an increase of 78 per cent in the number of journeys by road compared with 1962. This level of restraint might well be feasible without excessive congestion, using existing administrative techniques such as parking control, particularly in high density areas of business and shopping.

1.36 The third road plan was forecast to accommodate the total potential demand representing an increase of 100 per cent in the number of road journeys over 1962. This would cost over £2,000 million.

1.37 These results are the first quantitative indications of the effects of alternative levels of road investment and the associated restraint of demand. The results do not of themselves indicate what level of investment is most appropriate. They do, however, provide a valuable guide to the costs of serving the total demand for road space and various proportions of the total. They do enable some broad measures to be placed on the levels of restraint implied by different levels of investment so that the feasibility of achieving them can be assessed in the light of present techniques and experience.

The implications for public transport

1.38 The results of the studies provide two ways of evaluating the implications for public transport. In the first the results are examined before the necessity to restrain road demand has been taken into account. Although these results do not represent reality they do provide an opportunity to assess how the road plans affect public transport demand by direct competition. The second set of results apply to the more realistic situation in which road demand had been restrained and public transport demand increased as a consequence.

1.39 The results of the studies show that, before restraint is considered, the alternative road plans have very little effect on the demand for public transport. The reasons for this are to be found in the types of network tested and the characteristics of the demand. For the rail system providing the most suitable mode of travel for most work trips to the centre the road plans can offer little competition : no motorways penetrate an inner area of 47 square miles. The demand for rail services is affected more by the absolute levels of employment in the centre than the road plans. The planning forecasts for these tests envisaged an increase of 13 per cent in central area employment. To the extent that this may now not happen corresponding changes will occur in rail loadings and the effects of these changes in employment are likely to be more significant than the effects of the alternative road plans.

1.40 Most journeys by bus are short (80 per cent are less than 2 miles) and in the absence of restraint the main determinant of whether the trip continues to be made by bus is the existence of a car. The primary roads are too far apart for short trips so that variations in primary road networks have little effect on the number of bus trips.

1.41 When restraint is applied the number of trips made by bus and rail increases. The increase in public transport demand is not as large as the number of trips restrained from the road system since it is assumed that some journeys will not be made at all if they cannot be made by road. This is no more than the reverse of the situation that occurs when a family acquires a car and makes more journeys than were previously made when only public transport was available. The research indicated that when the level of restraint was 40 per cent 1·14 million additional trips might be made each day by public transport compared with no restraint. For a 6 per cent level of restraint 230,000 additional trips would be made by public transport. The research results do not take account of any possible effects that increased road congestion may have on the performance of bus services and the repercussions this might have on the number of bus passengers.

1.42 Comparisons between the public transport plans show that the effect of improved interchange in the centre is principally to improve the quality of public transport service. The changes in travel time were not sufficient to affect significantly the largely captive market. In so far as quality of service rather than travel time affects choice of mode the techniques have underestimated the potential demand for the better public transport system. However, the number of trips by private car to central London is comparatively small so that the maximum diversion possible is not large in proportion to total public transport loadings.

1.43 Opportunities to improve public transport by major new links attracting traffic from the roads are strictly limited, because of the extensive nature of the present public transport system. Improvements in service through increased speeds, relief of overcrowding and better interchanges will mainly benefit existing public transport users, either directly or through diversions from other public transport services. The effect on buses of improved rail or underground services is particularly marked.

The economic factors

1.44 A further important aspect of Phase III of the LTS was to define as many of the costs and benefits of the alternative transport plans as present knowledge and techniques allowed. It is unfortunately inevitable that many of the costs and benefits associated with a major social enterprise like a transport system cannot be quantified. Particularly important in this context are the effects of new roads and railways on the environment. Clearly these environmental effects can include both costs and benefits. A similar problem is encountered with restraint, which gives rise to both costs and benefits. The purpose of the economic analysis therefore was to place orders of magnitude on those costs and benefits that could be quantified so as at least to provide a basis for comparison for some aspects of the alternative plans.

1.45 Detailed studies were made to estimate the costs of providing the road systems and methods were developed that provide a basis for costing further road plans quickly and thereby avoiding the extensive and detailed work undertaken in Phase III. Although it was not possible to take full account of the environmental costs due to such factors as noise and visual intrusion it should be noted that the planning guidelines used (see Part 2 Appendix D) took some account of the environmental aspects including the cost of a wider land-take than is usual, the costs of forms of construction to minimise environmental effects and allowances equal to 10 per cent of the cost of construction for environmental improvements adjacent to elevated structures in residential areas. The information available from the public transport authorities did not permit such detailed cost estimates to be made for the public transport plans.

1.46 The benefits measured in the economic analysis are due to changes in the costs of making trips, the increased capacity of the system enabling more people to travel, and the additional tax paid by the generated traffic. To assess the changes in the cost of making trips both vehicle operating costs and time costs were taken into account. Time saved on business purposes was valued at the average wage rate and time saved on journeys to work and journeys for shopping and social purposes was valued at one third of the average wage rate. Some may view these as conservative estimates of the value of time and it is true that the overall results are sensitive to the assumptions made about both the absolute value and the relative value of different types of time. However these arguments are most relevant to the absolute quantities obtained in the analyses rather than the comparison of one plan with another.

1.47 Before any plans could be compared it was necessary to decide how the restraint of road traffic was to be achieved. The main results of the studies assumed that some effective and widely applicable form of control would be used so that the road plans being compared could operate efficiently without congestion. In these circumstances both the medium and high investment plans yielded one-year rates of return of about 10 per cent on the additional motorway cost over the low investment plan. The additional cost of the high investment plan over the medium plan yielded slightly less than 10 per cent. These rates of return apply to the forecast year and are difficult to compare directly with discounted cash flow estimates made by nationalised industries.

1.48 An estimate has also been made of the rate of return when congestion is the controlling factor in the operation of the networks rather than a restraint mechanism such as road pricing. In these circumstances the relatively greater congestion on the less expensive road system increases the benefits to be derived from construction of the high or medium investment system, and in this case the corresponding one-year rates of return are well over 20 per cent.

1.49 In reality it is likely that any practical method of restraint will have inherent costs and will be imperfect in operation and so fail to adjust traffic completely to capacity. Nor will congestion ever be allowed to rise to its worst levels. The most likely one-year rate of return of the large plans compared with the smaller road plan is therefore within the range 10–20 per cent.

1.50 The general conclusion to be drawn from the economic analysis is that the plans evaluated showed similar rates of return. This implies that if the actual level of road investment in London is in the range evaluated the economic returns are likely to be of the same order. Practical considerations of environment and finance indicate that actual road investment in London will be less than the medium plan evaluated in the research studies. The results of the economic evaluation confirm previous studies conducted by the government that there are significant social benefits to be derived from reducing congestion through control measures if practical and acceptable means can be devised.

Developing a plan for action

1.51 The planning, design, acquisition of land and construction of major new transport facilities can take up to ten years by the time all the public enquiries and investigations have been completed. In order to start this process for at least some of the new roads and also to provide a basis for local development plans in London it was important to establish a realistic programme which could be achieved within the likely limitations on finance and resources.

1.52 During 1967 the Council, in consultation with the Ministry of Transport, established likely investment limits during the 1970's. At the same time the results of the continuing research studies were becoming available to the Council. Of particular interest was the information on demand by area, the relative loading of different parts of the system and the detailed cost estimates. Studies also suggested that if budget limitations prevented the completion of a high capacity system such as the £1,650 million plan there were advantages in completing as much of the mileage as possible at a lower capacity and in this way establishing the pattern of facilities at the earliest date. With the same objective in mind the Council put higher priority on establishing new links than upgrading existing routes to motorway standard, such as the North Circular Road. The research studies had included and costed the North Circular Road in the medium and high investment plans as a road upgraded to dual four lane motorway standard.

1.53 In defining the programme for action the Council had to take into consideration the national motorway system that is in an advanced stage. The M1, M11, M12, M2, A20 improvements, M23, M3, and M4, were all committed schemes either on the ground, being constructed, or in final planning stages, and therefore likely to be constructed before any of the Council's new roads could start. The Council was also well advanced with the duplication of the Blackwall Tunnel and building its approach roads.

1.54 Given the pattern of demand, the relative loadings, the budget limitations and the need to fit in with committed schemes, the Council defined a network of rings and radials as an objective to be completed by the early 1980's. This network consisted of an inner ring and a new south circular road to form a complete middle ring with the existing North Circular Road which would be improved to a consistent standard but not made a motorway. The outer ring was included only in the north, west and southwest, where demands were highest.

1.55 The research studies indicated that dual four-lane motorways were generally required. However the Council's plan envisaged many sections of the network being initially constructed as dual two-lane motorways if detailed studies showed this to be feasible in each case. The primary roads included in the Council's ringway plan were estimated to cost £650 million including new Ministry of Transport roads but excluding both national and GLC roads that were programmed before 1971/72 (e.g. M11, M23 and the East Cross Route). The Council's ringway plan achieves 70 per cent of the mileage of primary roads evaluated in the medium investment plan but because of the sections initially to be constructed at half width has only about 55 per cent of the capacity.

1.56 The Council's objective has been evaluated using the procedures developed by the consultants, and the detailed results are discussed in Part 2. For comparison with the results given earlier it is estimated that the degree of restraint implied by this level of investment is between 25 and 30 per cent.

1.57 As with all studies of this kind the basic assumptions made during the study may have important repercussions on the results. A particularly significant factor in these studies was the capacity assumed for the existing road system. Capacity is difficult to measure precisely and can be strongly influenced by changes in social habits such as a spreading either in time or space of activity (longer shopping hours for example). In these studies the capacity of the existing system has been assumed in future to be such that it could accommodate an increase in demand over that observed in 1962 of about 30 per cent. In environmental terms it is clearly desirable to limit the increase of traffic on many existing roads but it is likely that a greater increase than 30 per cent has already occurred. But in the absence of greater primary road investment and/or stronger management measures, congestion is the limiting factor and existing roads are likely to carry more traffic until the conditions produced provide their own deterrent to movement. The using up of existing road capacity may have already exceeded the desirable level in many areas and can only continue for a limited period before congestion costs lead to a serious disruption of economic and social activity and further deterioration in environment.

The development of management measures

1.58 It is clear from what has been said previously that investment and management policies must proceed together. Once the broad investment possibilities had been defined it was possible to develop from the research studies an indication of the management measures necessary and how far these needed to be supported by legislation.

1.59 In a market where no direct prices exist, the management of demand for road space is limited to the less direct and less flexible methods of parking control. Road pricing is certainly not a short term possibility even if the technical, administrative and social difficulties of such a system can be solved.

1.60 Part 2 describes the study that was made as a basis for the development of a parking policy to cover not only the on-street situation, where progress was already being made, but also the supply and use of off-street facilities.

1.61 This revealed some important facts about the present parking situation and about future conditions. In particular it was shown that long-term parkers (over 4 hours), many of them commuters, accounted for one third of the vehicles parked but occupied almost 80 per cent of the parking capacity. The review of parking standards for new buildings showed that much of the space provided at the local authorities' insistence was used for commuting purposes. This anomaly arises from a policy on standards consistent with conditions before widespread control of on-street parking and discouragement of road commuters was contemplated. The studies showed that as the number of on-street spaces is reduced the local authorities with existing powers of control have been unable to prevent inroads by commuters into parking space intended to meet the needs of business, shopping, and residents.

1.62 A parking policy for London was developed with an emphasis on management so that parking space could be adjusted as investment in roads proceeded. This meant increasing the proportion of parking available to the public at large and reducing that in private buildings more in line with their essential requirements. In addition it was necessary to seek legislation to bring some limited measure of public control over the use made of public parking space. Finally a target figure was set for an increase in off-street parking space in central London that would cater for an increase in business, shopping and social journeys consistent with the developing capability of the road system but which involved a strong restraint on journeys to work by car in the centre.

The continuing development of policy and research

1.63 Once the broad investment plans and management objectives have been defined there is still a need to fill in more precise priorities, to study the impact of major roads on the environment and on the existing road system, and to review the policies in the light of changing circumstances in the urban region.

1.64 Part 2 describes the technical work that has been undertaken, and that is continuing, to give guidance on the priorities to be afforded different parts of the primary road system. An early conclusion of this work was that there were significant benefits to road users to be derived by completing the southern half of Ringway 2 at an early date. This is because of the nature of the pattern of primary roads already committed, the pattern of demand and the complete absence of any existing orbital road in south London.

1.65 The Council's objective on primary roads involving an expenditure of £650 million recognised the need to spend a further £210 million on the existing road system. As this report is prepared the findings of the studies reported here are being used to study the secondary road problems in the light of the primary road policy.

1.66 Part 2 also describes a study being undertaken to evaluate the local environmental effects of the levels of traffic forecast for 1981 associated with the development of the primary road network. One of its main purposes is to assess the extent to which the removal of large quantities of through traffic would bring opportunities to improve the environmental conditions of the area as a whole by reduction of, for example, noise and hazards to pedestrians. The study has advanced understanding and knowledge, particularly in the measurement of environmental factors. It has already shown that for the area selected the projected motorway and secondary routes will divert a substantial proportion of traffic from the local streets. However, it also indicates that much of this could be replaced by locally generated traffic which is expected to increase as car ownership rises. Thus, total traffic volumes within the area could be similar to those of today. The study demonstrates that in this sample area local volumes of traffic can be reduced by a combination of additions to the system, such as new motorways, and parallel local management measures. If the new motorways were not built, management measures to alleviate conditions in the area by diverting through traffic to other secondary roads would be impracticable because of the lack of adjacent less congested streets. However, once the primary road exists local options can be encouraged so that traffic measures that primarily inconvenience local residents and local traders can be judged more directly in terms of local benefits. This study covers only a limited area and its conclusions may not be of general validity but it illustrates the need to define, in cooperation with the local authority concerned, the parallel management needs of an area as the related primary network proposals become commitments.

1.67 A continuing need will always be to keep strategic policies under review in the light of changing circumstances. This must not be used as an excuse for delaying necessary action but equally the sensitivity of major decisions to the planning context should be assessed. It has been indicated earlier that the economic factors are the most significant in terms of the amounts of future travel expected. Changes in population and employment will have their local effects but will not significantly alter the basic case for a primary road system in London. Part 2 describes some general sensitivity tests that assumed major reallocations of employment and population in London in order to test the validity of the primary road concept in different planning situations. In addition, Part 2 gives results for revised travel forecasts for London based on a 1·5 per cent decline in population and a 2 per cent increase in employment for the GLC area compared with 1961 in contrast to the previous forecasts based on increases of 2·5 per cent and 11 per cent respectively. Also, and most significant, annual income growth was assumed to be 2·3 per cent compared with 2·8 per cent previously (18 per cent overall decrease). The net effect of these changes is shown to be a 6 per cent decrease in total travel and 6·5 per cent decrease in the demand for journeys by road in 1981 compared with the forecasts made in the LTS.

1.68 Since these revised forecasts were made, the 1966 census material and further studies for the GLDP have been completed and these indicate the possibility of further decreases in population and employment by the early 1980's, perhaps by more than 10 per cent compared with 1961. The detailed effects of these new estimates are being evaluated but it is clear from the work already reported and from the degrees of restraint implied by current investment programmes that there is no possibility of present strategic policies supplying an excess of road capacity compared with the potential demand. If the trends are realised in practice the general easing of pressures may provide better opportunities to help the overall environment of London than appears feasible today.

Conclusions

1.69 The purpose of this report is to demonstrate how transport planning studies have been related to the development of policy by the GLC and by the LCC before it. Since the responsibility of both authorities in transport matters has hitherto been mainly concerned with the development and management of the road system this has naturally been the main concern of the studies; but as this report shows the scope has been wide enough to ensure that the needs have been assessed as thoroughly as techniques permit, that investment and management policies have been consistently related, and that the capabilities of public and private transport to meet every demand have been objectively evaluated.

1.70 The research studies alone do not, and could not, demonstrate that there is one level of investment in transport that is 'right'. Priorities between roads, houses, railways, schools and open spaces must depend on more than computer output. In fact the security of the strategic concepts now advanced depends as much on local planning feedback as on continued strategic studies.

1.71 Planning in London is making a further step forward as the GLC is given overall transport planning responsibilities. The comprehensive and wide-ranging nature of the research studies conducted during the last five years has provided a sound and practical base for these new responsibilities to be carried out as part of a continuous process.

Part 2

Detailed studies of the factors influencing travel demand were first attempted in the United States in the 1950's. They were comprehensive in that they assessed the effect of population and land use characteristics on trip making patterns and were able to forecast future trip making over a long period and for a wide area. In 1960 a comprehensive study of road traffic was initiated by the London County Council and the Ministry of Transport. A joint firm of British and American consultants, Freeman Fox, Wilbur Smith and Associates, were appointed to carry out the study.

In this part of the report the London Transportation Study (LTS) is described followed by a description of related work carried out by the Greater London Council. Phases I and II of the LTS have been published previously in 1964 and 1966 respectively. But for convenience the main facts from these phases important to the understanding of Phase III, the final and most recently completed part of the study, are summarised in Chapters 2 and 3.

Phase I covered the collection of basic data in 1962. At first the study was concerned only with private transport within the London County Council area, but the necessity for a comprehensive approach to transport within the London conurbation was soon realised and the terms of reference broadened. It was decided to extend considerably the boundaries of the study area and to gather data from non-car-owners.

Several other changes in the scope and method of the study have since been made. A number of lessons have been learned and new methods developed, a consequence of the fact that this was the first such study to be undertaken in this country and one of the largest to be undertaken anywhere in the world. In the course of its evolution changes took place in the powers and responsibilities of local government in London. The formation of the Greater London Council occurred after the completion of Phase I. An important responsibility of the GLC is the preparation of a Greater London Development Plan. So one of the purposes of Phases II and III of the LTS has been to provide a background for the transport plan included in the GLDP.

In Phase II forecasts of planning factors were made for 1971 and 1981, which were taken as the target years for the study. Possible future transport networks for these years were devised, incorporating new high speed road links. Statistical analysis of the relationship of personal characteristics to travel patterns in 1962 was carried out in order to produce a mathematical model. The model was then used in conjunction with forecasts of the planning factors and possible transport networks to estimate the future potential amount and pattern of travel and the adequacy of the various sections of these networks.

In Phase III a number of the stages of Phase II were repeated, incorporating modifications as a result of Phase II experience. Future road and public transport networks were designed and costed. The mathematical model was considerably refined and used to estimate not only potential trip making but patterns of trip making possible when the limitations of transport capacity are taken into account. After assignment of these trips to the possible transport networks an assessment was made of the economic and operational consequences of implementing each of the transport plans. The most recent forecasts of the future population and employment in Greater London are lower than those used in the LTS which were the best which could be obtained when they were assembled. Revised estimates are being used by the Greater London Council in their latest estimates of future travel. It is worth noting, however, that the changes in the population forecasts, though important, are very much smaller than forecast changes in the other major determinants of the quantities of travel namely income and car ownership.

The last chapter of the report describes work carried out by the Council in parallel with or subsequent to the LTS. This includes work on forecasting such as the calculation of car ownership figures and peak hour factors as well as separate studies of parking and the effect of traffic on the environment. A section is also included which describes the results of an evaluation of the Council's London Roads Objective network of primary roads and a study of the most appropriate order of construction of several major projects within it. Finally a number of technical appendices are included.

It is worthwhile noting that not the least benefit of undertaking a study such as the LTS has been the greatly increased understanding of the transportation problem acquired by a wide range of authorities and individuals. Research organisations and universities have also been stimulated to work within the field of transportation forecasting and economic analysis thus furthering the understanding of the total process and providing an even better basis for future decisions.

2

Figure 2.1 Stages of the London Transportation Study

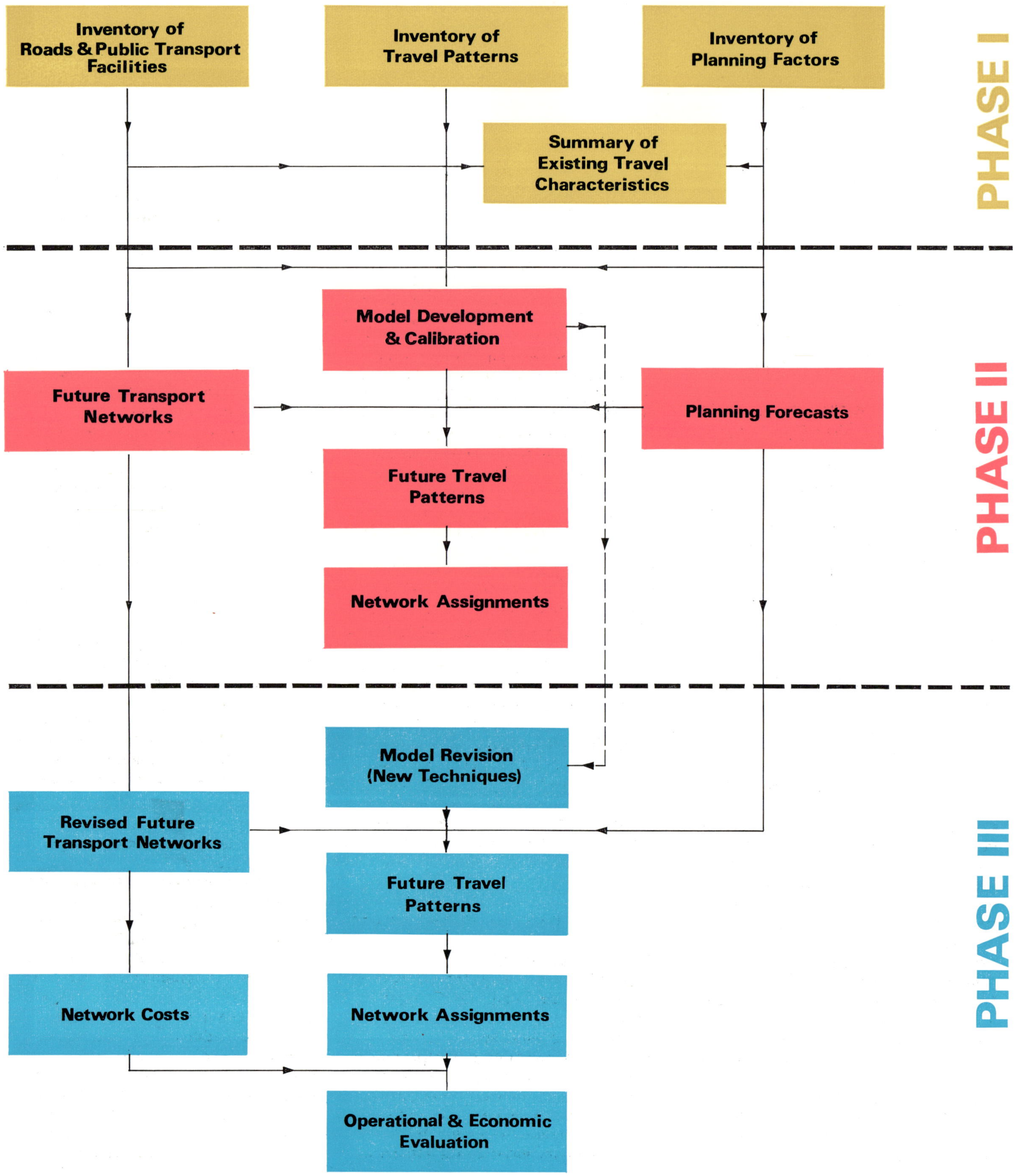

Chapter 2 Basic relationships in 1962 (Phase I)

The approach adopted

2.1 The main stages adopted in the three phases of the LTS are shown in Figure 2.1 from the initial survey of trip making and related factors to the evaluation of proposed measures to improve transport in London. Only the main stages are shown, the procedures adopted and the results obtained being discussed in the following sections of the report.

2.2 In Phase I the characteristics of the travel pattern and the means of travel in 1962 were investigated. At the same time a survey was made of information about the population such as income and car ownership and data on the distribution of land uses, residences and work places. As a result it was possible in Phase II to establish relationships between person or household characteristics and their travel characteristics. These relationships can be expressed in a mathematical form and then used with forecasts of population and land use characteristics to estimate future travel patterns on the basis of given assumptions.

2.3 Thus the approach is to develop basic behavioural relationships which may be expected to remain fairly constant over time and not simply to project trends in travel quantities and patterns for which the original data might quickly become out of date. By revising the assumptions about, for example, population and income in the future, the forecast travel patterns can be kept up to date. It is still necessary to check the basic relationships, but not frequently. Finally the approach is directed not at finding the exact forecast of what will happen in a given year in the future, because this depends on the correctness of the assumptions and the policies adopted in the meantime, but rather at evaluating alternative courses of action or assessing the implications of a given course of action.

Data collection

General

2.4 For a study of this nature a wide range of data is required. Discussing these requirements in terms of the process illustrated in Figure 2.1, the inventory of roads and public transport facilities included a study of road capacities and of the speeds and volumes of traffic. The study of rail facilities included an examination of train utilisation, track capacities, and the isolation of critical points, but was not as detailed as the road studies since the purposes of the study at this stage did not require it. The inventory of travel patterns on a typical weekday required detailed knowledge of all movements made inside the LTS area by people living both within and outside the area by all modes of travel, including bus and rail, car, motorcycle, cycle, coaches, taxis, and by commercial vehicles. Besides the mode of travel, information was obtained on the time, purpose, location and land use at the origin and destination of each trip. The inventory of planning factors included population, numbers in households, number of employed residents in households, the income and car ownership of households and more generalized factors such as residential density and distance from the Central Area.

Travel patterns—the home interview

2.5 The great majority of travel in the LTS area is made by members of households living within the area. By interviewing people at their homes it was possible to obtain a record of all the trips made by members of sample households on a particular day and to relate these trips to the economic background of the people concerned. There was of course some under-reporting, but by careful interviewing this was minimised.

2.6 The area covered by the LTS is shown in Figure 2.2 and has an area of 941 square miles, approximately half as large again as the Greater London Council area which was established three years after the survey was carried out. In an area containing nearly three million households it was clearly not possible to interview at every household and therefore a sampling procedure had to be adopted. The size of the sample selected was 3·6 per cent of households which owned cars and 0·6 per cent of those which did not own cars in 1962. The selection of sample size is more fully discussed in Volume I of the LTS, but the car-owning and non-car-owning households were sampled separately because it was originally intended to confine the study to car-owners alone. As a consequence of the low proportion of non-car-owners interviewed, the quality of information for certain trips is low, for instance for bus journeys, most of which are made by non-car-owners. This limits the value of the results which are based on this information. It is important to note, however, that many trips by public transport are made by people from car-owning households, particularly journeys to work, so that the overall sample size for public transport trips lies between 0·6 and 3·6 per cent.

Travel patterns—other surveys

2.7 Separate surveys were made of the movements of taxis and commercial vehicles garaged within the LTS area. Drivers of 20 per cent and 15 per cent respectively of these vehicles kept a log book of all movements made in one day.

2.8 In order to obtain information about trips within the LTS area made by non-residents of the area and commercial vehicles not covered in the internal survey, an external cordon survey was carried out. This involved interviewing a sample of inbound drivers crossing the boundary of the LTS area and a

Figure 2.2 Internal traffic districts and sectors (Phase I)

London Traffic Survey boundary
Traffic sector boundary
Traffic district boundary
River Thames
Greater London Council boundary (1965)
Traffic sector number 2
Traffic district number 147.

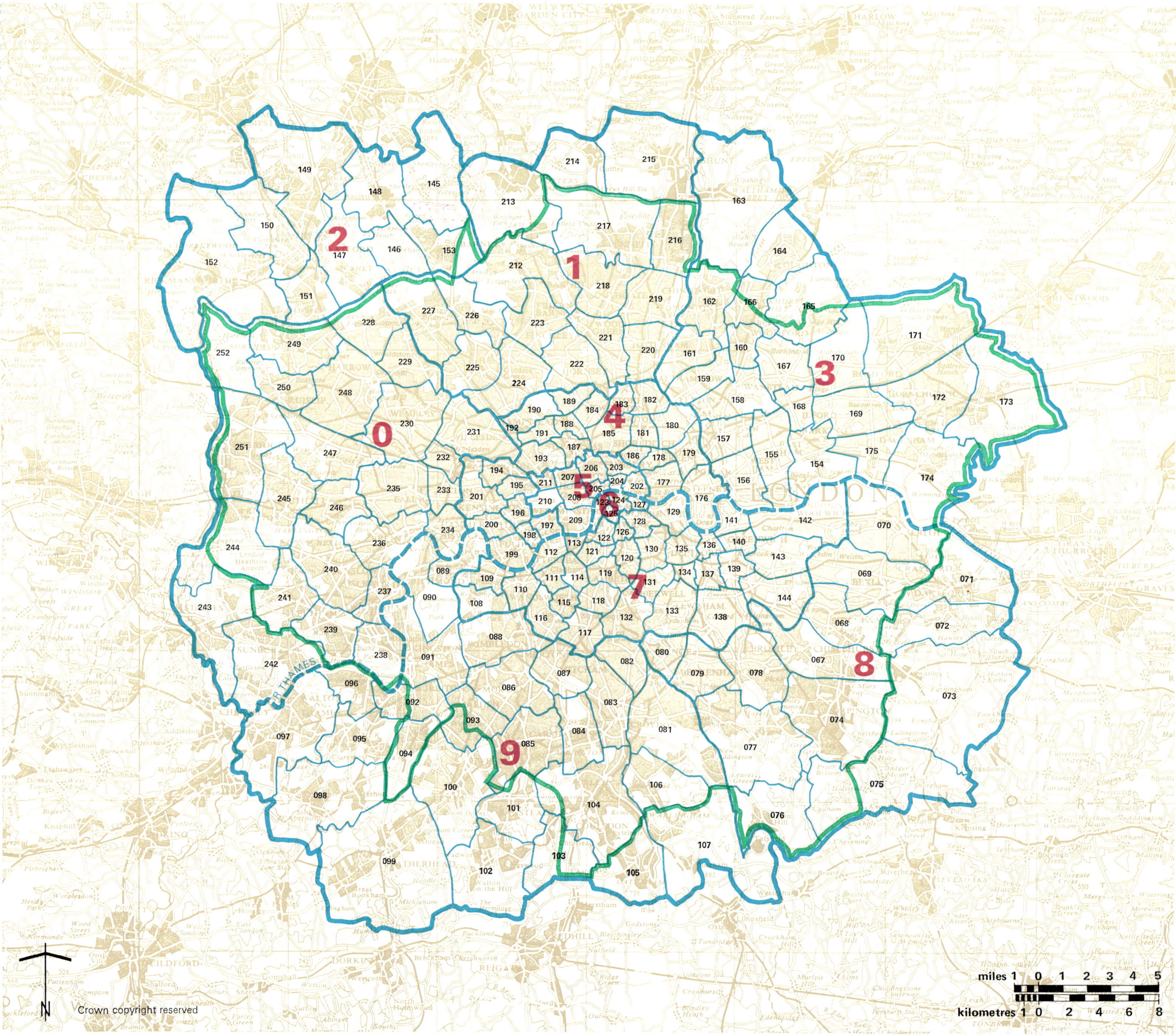

postcard questionnaire of passengers on buses and trains crossing the cordon.

Other data requirements

2.9 The demographic data required for the study such as population, household size, employment, car ownership and income was obtained partly from the home interview and partly from supplementary sources, in particular the 1961 National Census. Comparison of census and home interview data indicated that there was good agreement between the two.

2.10 Inventories of road and public transport facilities in 1962 were made, including particularly the capacity, volumes and speeds of links.

Coding

2.11 The first stage in the data analysis process was to divide the area into zones, districts and sectors. There are 933 zones, 186 districts and 10 sectors for specifying the location of trip origins and destinations with different degrees of precision. These have been the basis for most subsequent work including further public transport surveys and are also being used in the continuing programme of transportation studies. Figure 2.2 illustrates the district and sector zoning. The area outside the LTS area was also divided into zones for analysis purposes.

Planning data

Population

2.12 The LTS area contains almost 17 per cent of the population of Great Britain and the rest of the South-East contains a further 17 per cent. The population distribution is illustrated in Tables 2.1 and 2.2. Figure 2.3 shows the 1962 distribution of net residential density within the LTS area and clearly shows the denser population towards the centre. The area of high residential density of over 65 persons per acre occurs largely within the former LCC boundary. It is particularly in this area that increases in car ownership in terms of movement and parking will have the greatest impact.

Table 2.1 Distribution of households and population in households by sector (Phase I)

Traffic sector	Area (sq miles)	Population in households		Households		Gross population density (persons per sq. mile)	Average number of persons per household
		Number	Percentage	Number	Percentage		
0	151·75	1,347,980	15·9	448,710	15·2	8,900	3·00
1	107·98	941,090	11·1	320,870	10·9	8,700	2·93
2	80·72	220,690	2·6	69,330	2·3	2,700	3·18
3	140·67	1,229,330	14·5	401,870	13·6	8,700	3·06
4	40·49	1,370,290	16·1	536,980	18·1	33,800	2·55
5	9·03	190,790	2·2	86,000	2·9	21,100	2·22
6	1·40	30,140	0·3	11,280	0·4	21,500	2·67
7	65·39	1,432,150	16·8	500,090	16·9	21,900	2·86
8	132·76	581,300	6·8	190,660	6·4	4,400	3·05
9	210·85	1,161,780	13·7	393,220	13·3	5,500	2·96
Subtotal central area	10·43	220,930	2·5	97,280	3·3	21,200	2·27
Total LTS area	941·04	8,505,540	100·0	2,959,010	100·0	9,000	2·87

Figure 2.3 Net residential density by traffic district (Phase I)

Persons per net residential acre

0–24		45–64	
25–34		65–109	
35–44		110 and over	

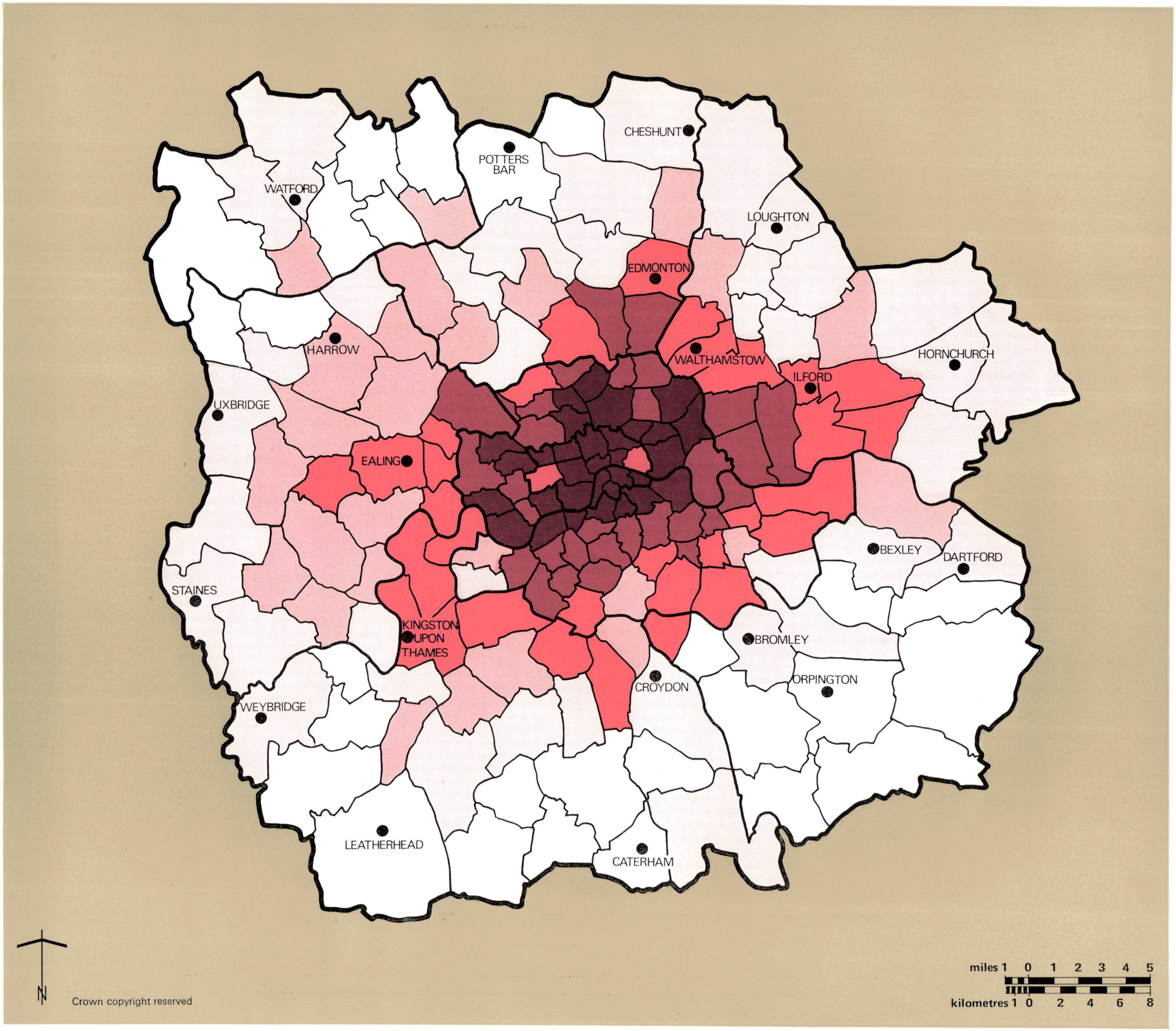

Table 2.2 Population of the LTS area and other significant planning areas (Phase I)

Planning area	*1961 population*	*Percentage of population in Great Britain*
LTS area:		
Private households	8,505,540	16·6
Other	321,080	0·6
Total LTS area	8,826,620	17·2
Remainder of London Metropolitan Region	3,639,320	7·1
Total London Metropolitan Region	12,465,940	24·3
Remainder of South-East England	5,200,320	10·1
Total South-East England	17,666,260	34·4
Remainder of Great Britain	33,617,590	65·6
Total Great Britain	51,283,850	100·0

Employment

2.13 Employment distribution is important to transport planning because of the large number of trips which are to or from work. Figure 2.4 illustrates the density of employment in the LTS area, and shows again the intense concentration in central London. The distribution of employment within the GLC area by industrial group is shown in Table 2.3. The bulk of the employment is in offices and manufacturing and most of the remainder occurs in the retailing and wholesaling trades. Figure 2.5 illustrates the distribution of employed residents in the LTS area. This is important because the number of trips generated depends on the number of employed residents in the household.

Car ownership

2.14 The distribution of car-owning households throughout the London area is a vital factor in transport planning and is illustrated in Figure 2.6 and Table 2.4. What is important to note from Figure 2.6

Table 2.3 Distribution of employment by sector and industrial class group (Phase I)

Traffic sector	*Number of employees by industrial class group*					
	Retailing	*Manufacturing*	*Wholesaling*	*Other**	*Total*	*Percentage*
0	55,240	282,690	17,420	284,100	639,450	13·4
1	40,090	148,140	10,110	169,620	367,960	7·7
2	8,370	45,010	1,990	45,440	100,810	2·1
3	43,760	224,890	12,170	179,140	459,960	9·6
4	67,300	219,990	34,160	340,430	661,880	13·9
5	86,570	314,840	109,890	827,720	1,339,020	28·1
6	6,080	26,480	16,310	56,100	104,970	2·2
7	62,600	159,280	18,180	264,050	504,110	10·6
8	21,230	76,490	3,370	68,150	169,240	3·5
9	55,420	169,600	10,310	189,820	425,150	8·9
Subtotal central area	92,650	341,320	126,200	883,820	1,443,990	30·3
Total LTS area	446,660	1,667,410	233,910	2,424,570	4,772,550	100·0
Percentage	9·4	34·9	4·9	50·8	100·0	—

*Extractive, construction, utilities and service industries

Figure 2.4 Employment density by traffic district (Phase I)

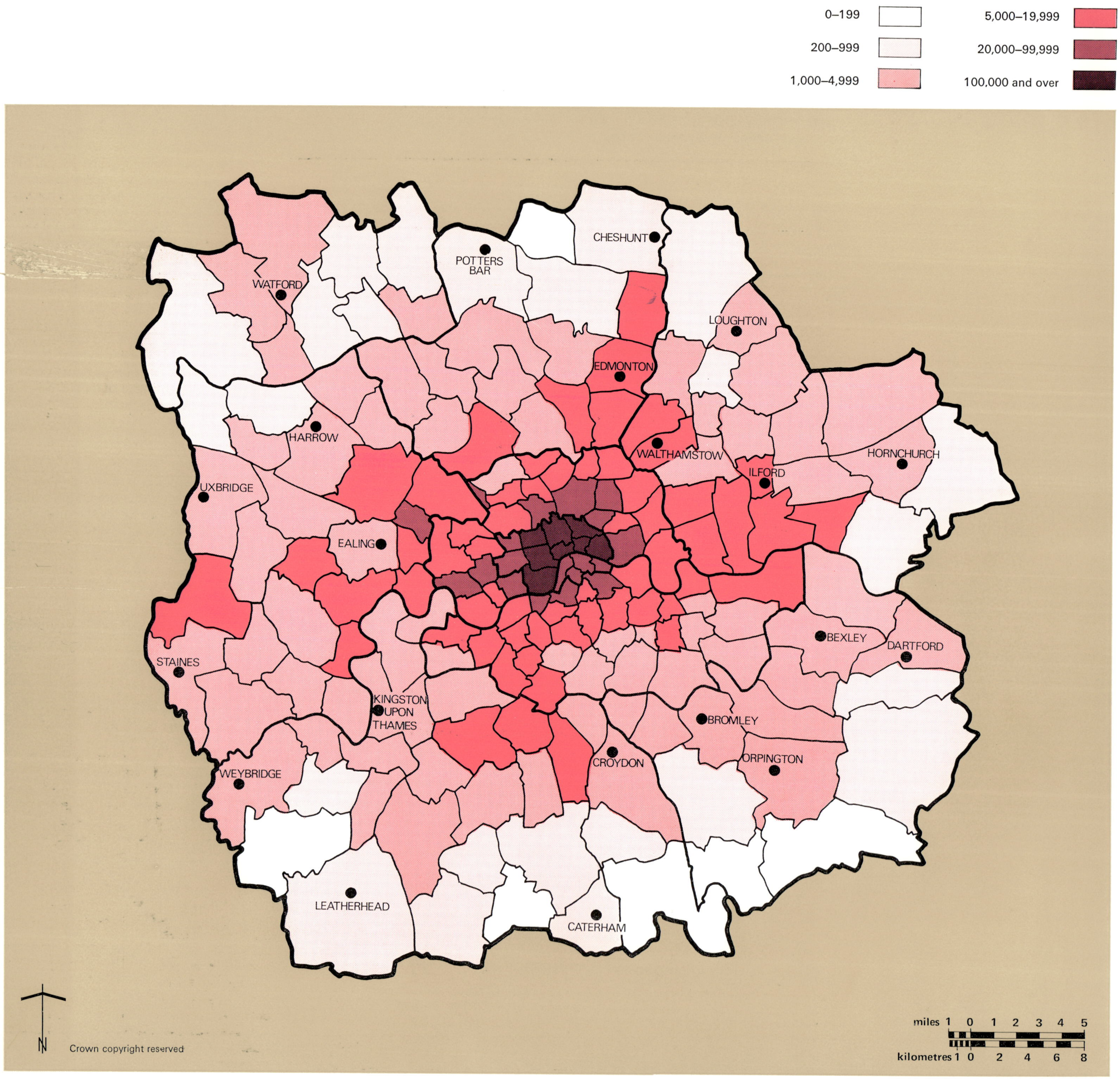

Table 2.4 Car ownership (Phase I)

Traffic sector	*Percentage of households* Car-owning	Non-car-owning	*Number of car-owning households* 1-car	2-car	3-car or more	Total	Number of cars	Number of cars per car-owning households
0	44·1	55·9	183,984	19,601	2,098	205,683	229,620	1·12
1	43·7	56·3	116,652	14,883	1,815	133,350	152,036	1·14
2	48·7	51·3	28,957	3,967	592	33,516	38,713	1·15
3	38·4	61·6	141,937	11,063	1,061	154,061	167,366	1·09
4 & 5	26·9	73·1	152,109	13,189	1,856	167,154	184,443	1·10
6 & 7	29·2	70·8	138,079	10,765	853	149,697	162,235	1·08
8	46·4	53·6	79,191	8,232	717	88,140	97,928	1·11
9	48·5	51·5	166,572	21,664	2,327	190,563	217,112	1·14
Total LTS area	37·9	62·1	1,007,481	103,364	11,319	1,122,164	1,249,453	1·11

Table 2.5 Distribution of LTS area households by household income class (Phase I)

Income class	*Range of total annual household income*	*Number of households*	*Percentage*
1	Under £500	413,700	14·0
2	£500–£999	1,174,500	39·7
3	£1,000–£1,499	817,400	27·6
4	£1,500–£1,999	316,300	10·7
5	£2,000–£2,999	157,700	5·3
6	£3,000 and over	79,400	2·7
	Total	2,959,000	100·0

is the generally decreasing car ownership rates towards the centre of London, with the exception of parts of the West End. The areas with low car ownership at present tend to be those where population densities are greatest. It is in these areas that car ownership growth has the most potential, though constrained somewhat by the high densities.

Income

2.15 A knowledge of income distribution is important because increasing incomes in future may be expected substantially to increase both car ownership and the demand for travel by all modes. Table 2.5 shows the distribution of incomes and the geographical distribution of incomes is shown in Figure 2.7. Areas of high income in 1962 correspond closely to those areas with high car ownership.

Figure 2.5 Employed residents by traffic district (Phase I)

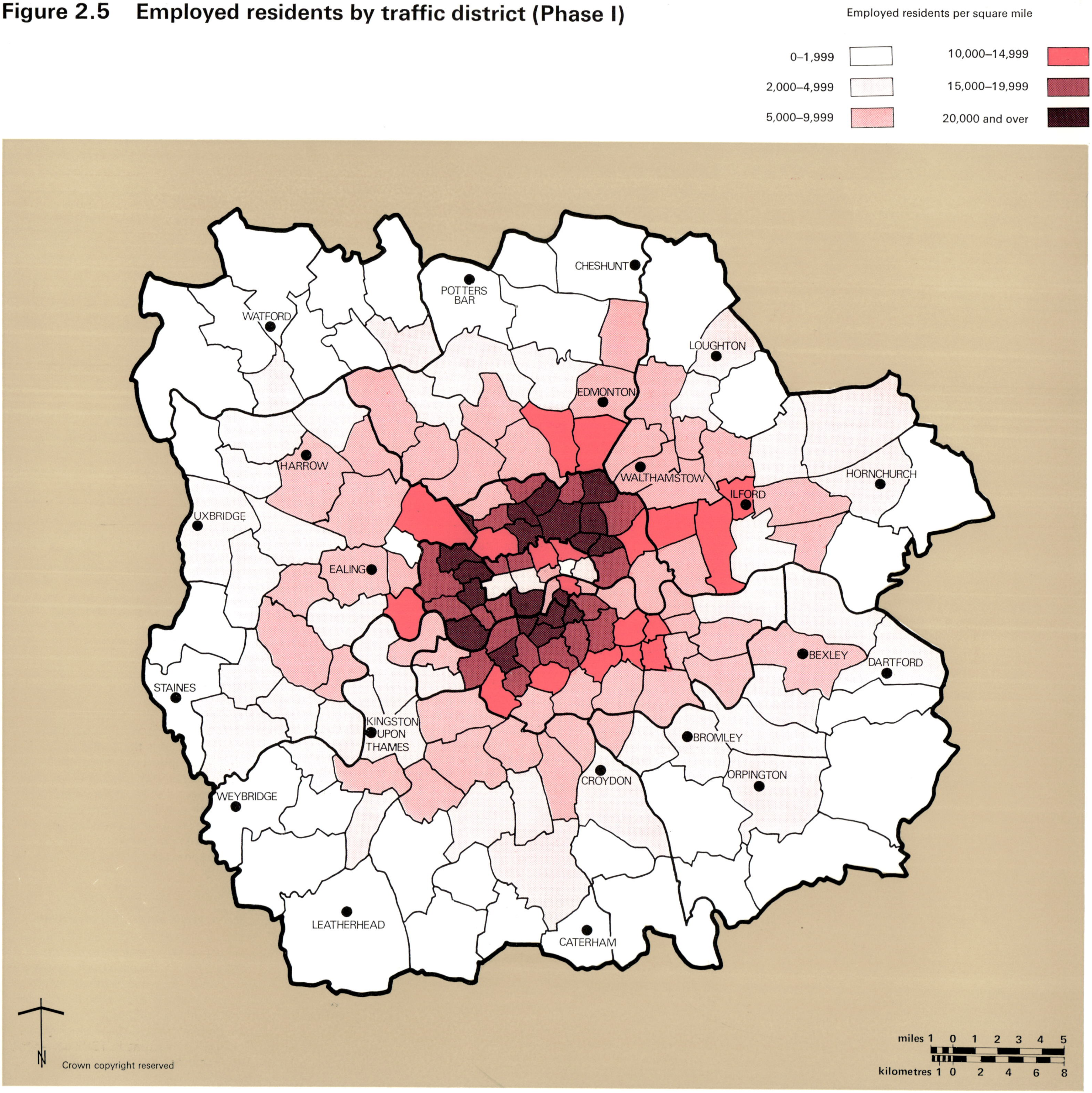

Figure 2.6 Car ownership by traffic district (Phase I)

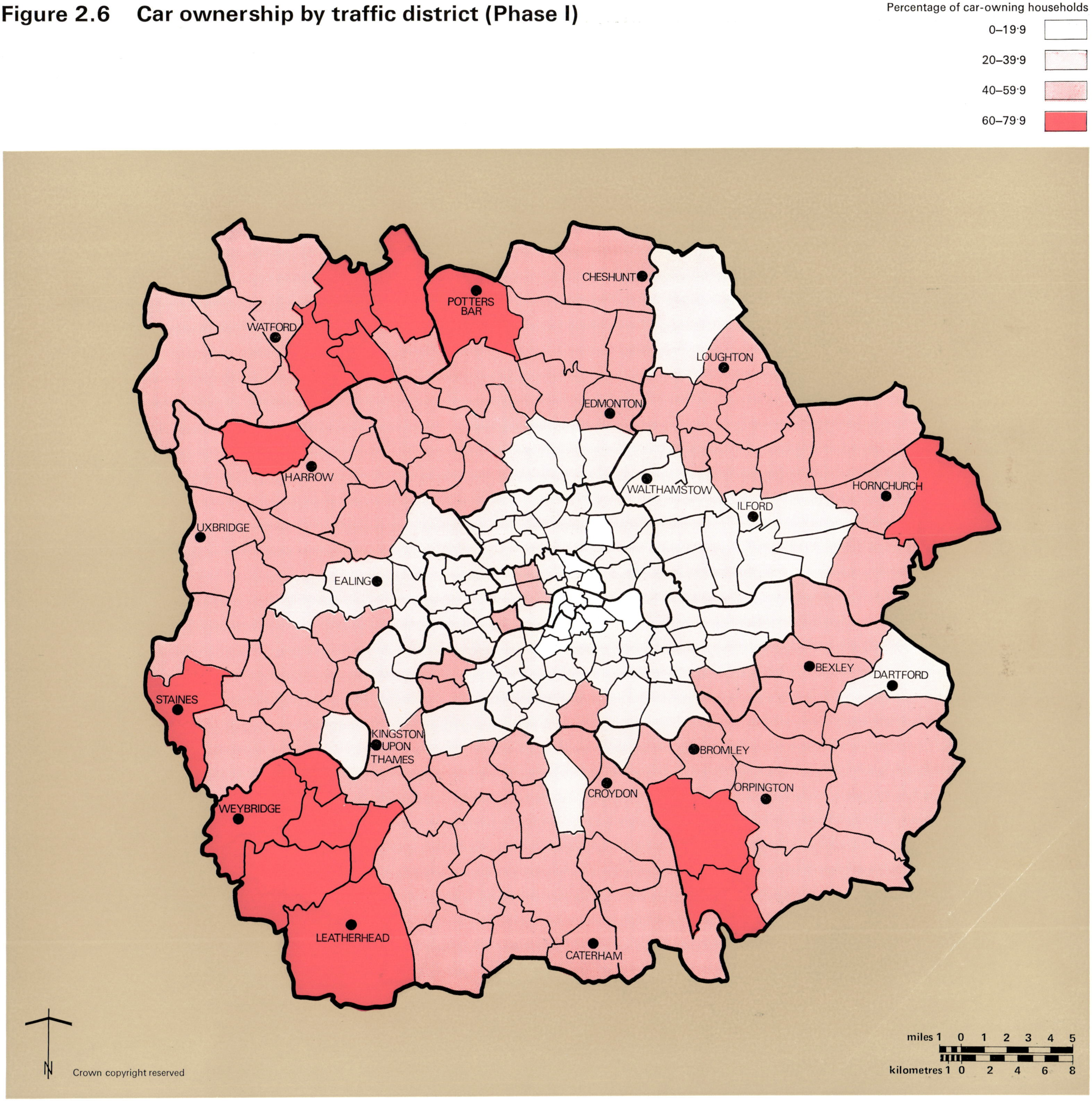

Figure 2.7 Mean household income by traffic district (Phase I)

Mean household income (£)	
0–999	
1,000–1,499	
1,500–1,999	
2,000–2,499	

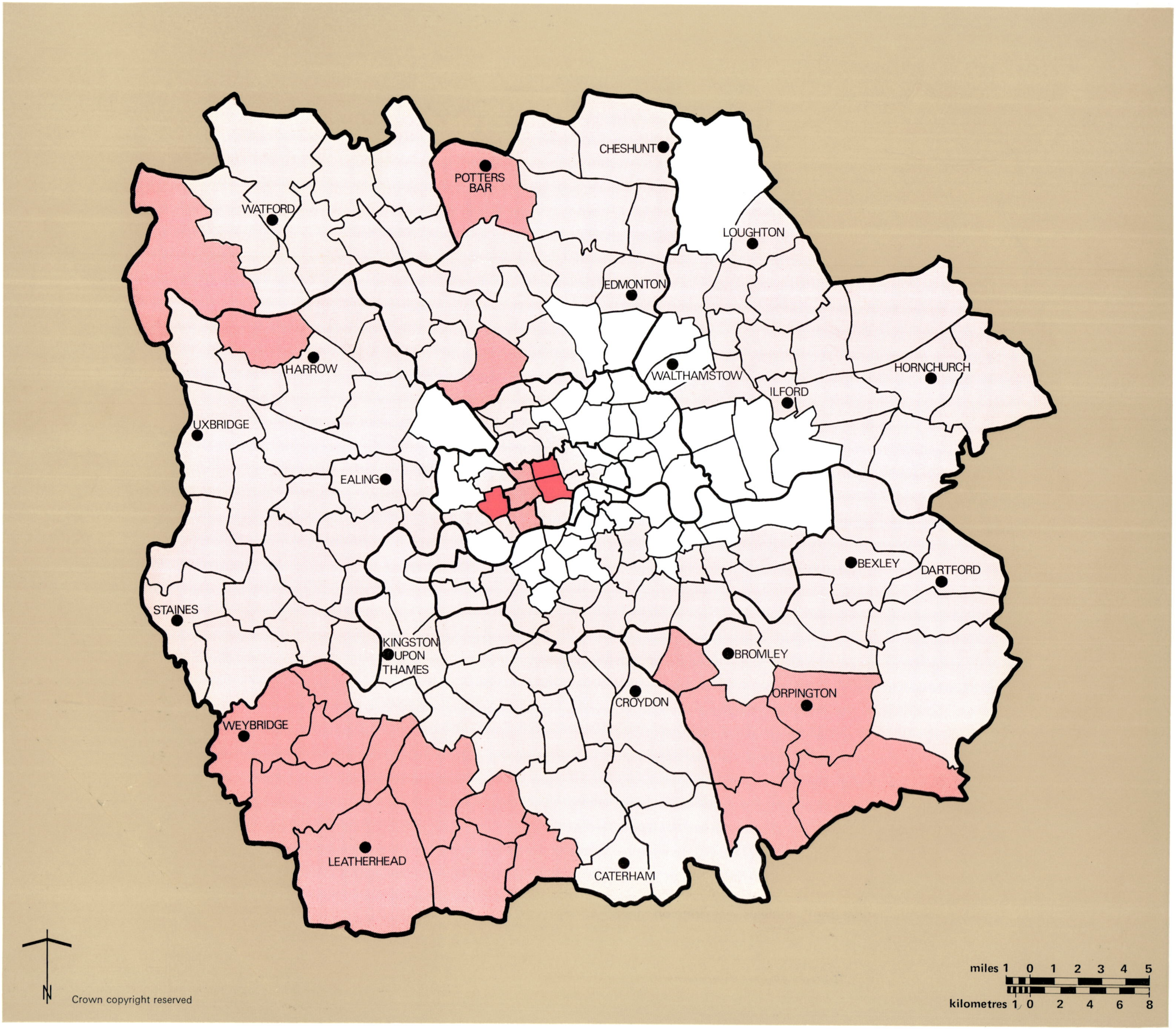

Travel characteristics

Basic relationships

2.16 Throughout the LTS the word trip means the one-way movement of a person or commercial vehicle between an origin and destination (generation and attraction in strict technical terms) and, unless specified otherwise, the analysis is restricted to movement by road vehicle or public transport. Figures 2.8 to 2.11 show some of the factors which affect the amount of travel by members of a household, namely income, residential density and employment.

2.17 Income is the most important factor determining future car ownership levels and Figure 2.8 shows how car ownership increases with household income. Figure 2.9 shows perhaps the most significant fact of all for transport planning, that car-owning households in 1962 made many more trips than non-car-owning households. On average when a household acquires a car its rate of trip making approximately doubles. This is an important fact to be borne in mind in assessing the future level of travel demand. Figure 2.10 shows that as the residential density increases, trip making decreases for both car-owning and non-car-owning households. This is due to a number of causes among them being that people in high density areas tend to be from the lower income groups and that due to congestion there is generally less freedom of movement. In addition, a greater proportion of trips can be made conveniently by foot. Figure 2.11 illustrates the effect of household composition on trip generation. As the number of employed residents in a household increases so naturally do the work trips increase.

Trip purpose and mode

2.18 Information on 1962 person trips was divided into 12 categories of trip purpose, but in estimating future trips most of the analysis was done on the basis of three categories only:

a. Work

b. Other home-based

c. Non-home-based

Figure 2.8 Effect of household income on car ownership (Phase I)

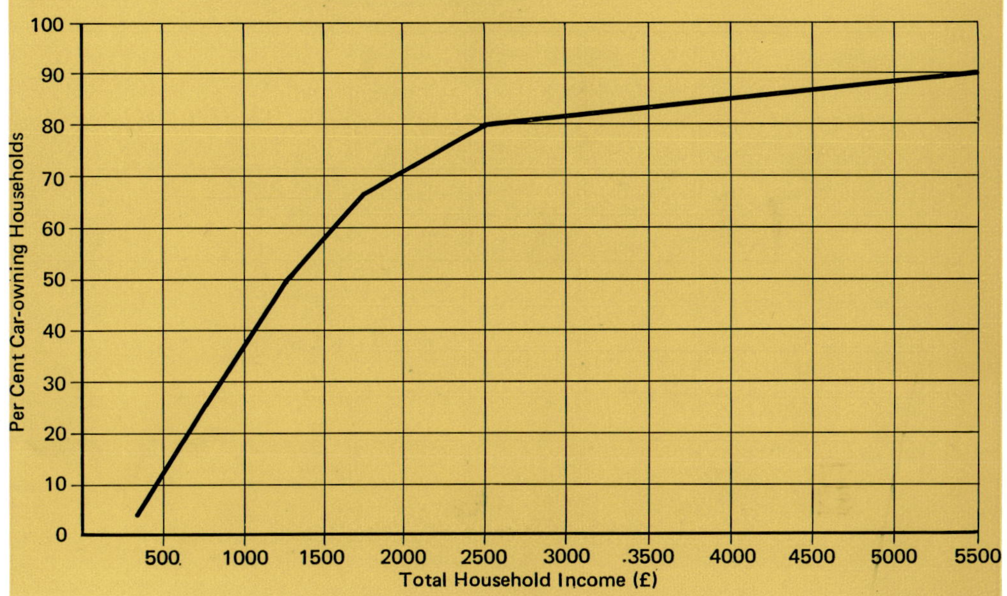

Figure 2.9 Effect of household income on trip generation (Phase I)

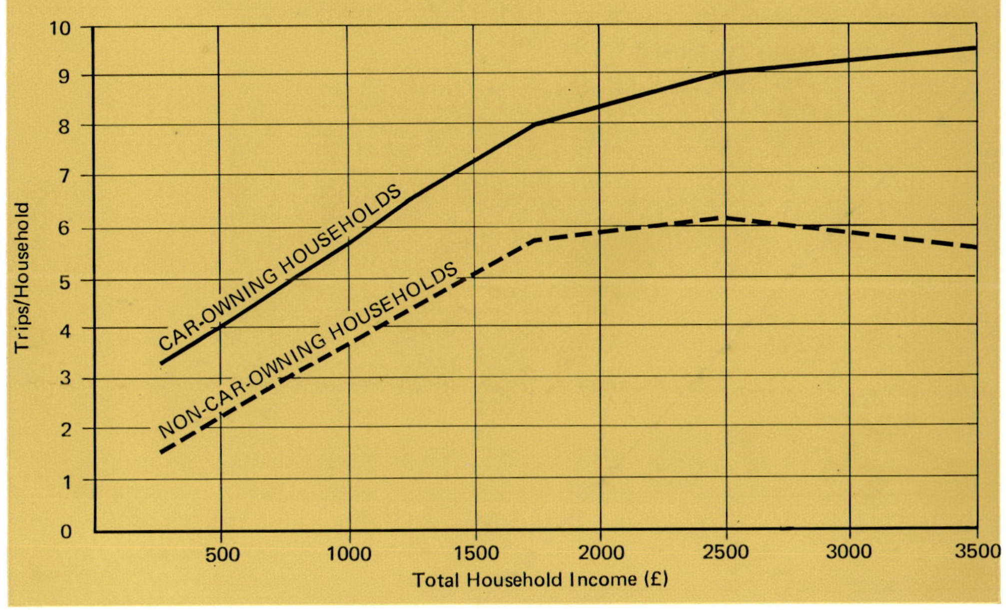

Figure 2.10 Effect of residential density on trip generation (Phase I)

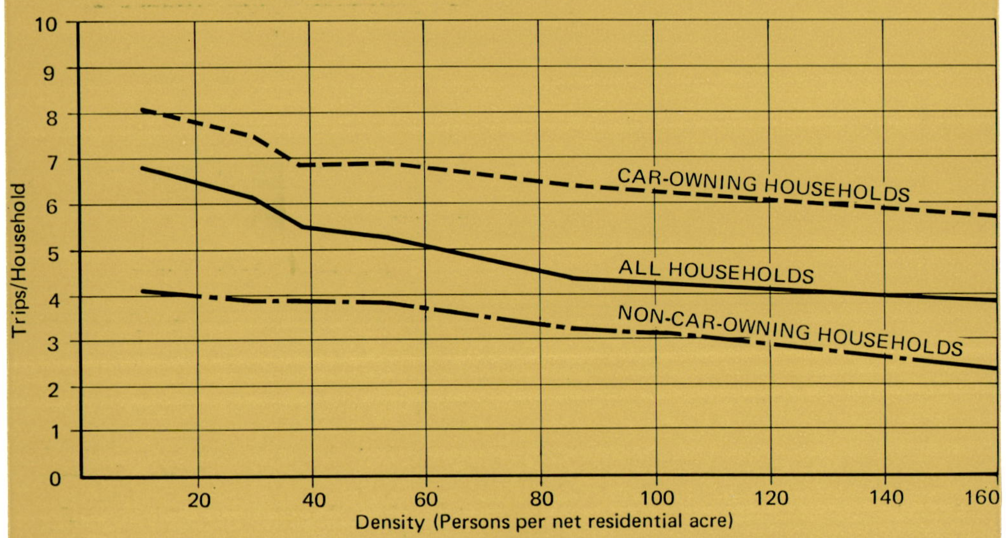

Figure 2.11 Effect of employed residents on trip generation (Phase I)

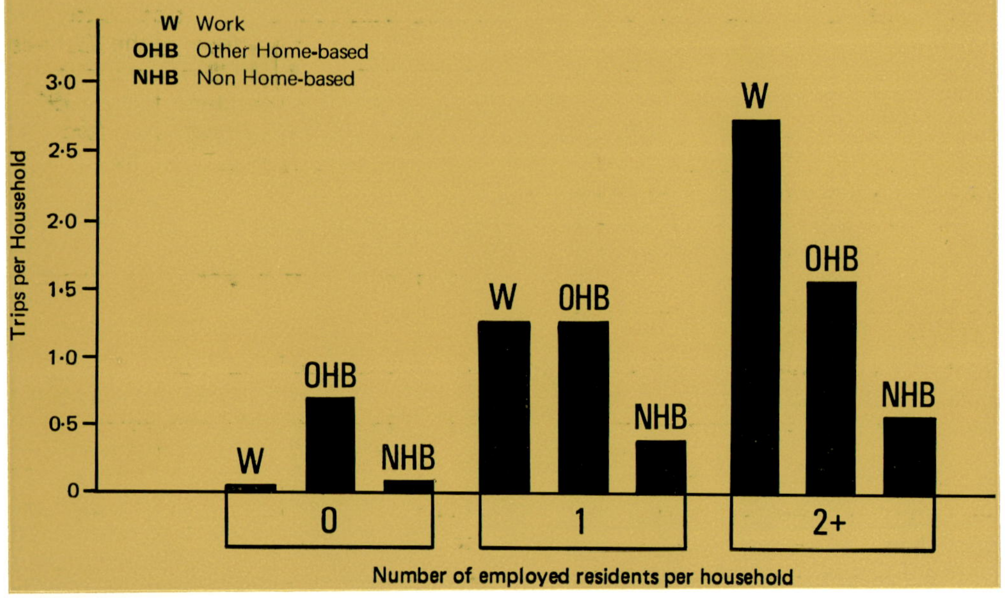

The first category consists of trips in both directions between home and work, other than trips on business, and accounts for almost half of all trips in 1962. The second category consists of trips in both directions between home and some purpose other than work, shopping for example, and accounts for a further 40 per cent. The third consists of trips which neither begin nor end at home, trips from work to shop for instance. This three-way classification is the basis of the forecast of future trips in relation to the distribution of homes and workplaces.

2.19 In considering the mode or means of travel technical difficulties are encountered with existing methods of analysis when considering trips which use more than one mode. In the survey of travel, trips were analysed on the basis of all modes used in the course of a trip. However, for the purposes of forecasting, it is necessary to classify trips by 'main mode', that is the mode used for the longest distance. This means that for the future it is not possible to obtain a direct forecast of individual journey link totals, for example the number of people using the bus to their final destination in central London after a long rail journey.

Trip generation

2.20 Approximately 90 per cent of all trips have either their origin or their destination at home. For such trips the home end is defined as the 'generation'. Hence, the amount of trip generation can be related to household characteristics and the rate of trip generation in future can be estimated by forecasting the characteristics of future households. At this stage of the work it was not possible to take account of the possible effects on trip generation rates of *increases* in congestion, parking charges or improvements in public transport beyond those conditions prevailing in 1962. From the analysis of household characteristics it was found that the most important factors which influence trip generation rates are car-ownership, number of employed persons, income and residential density. These were each divided into three ranges or categories so that each household fell into one of

Table 2.6 Trip generation rates* (Phase I)

Income class	No car	(change)	1-car	(change)	2-car or more
Low income	2·06	+1·99	4·05		
(change to medium)	+1·01		+1·24		
Medium income	3·07	+2·22	5·29	+2·32	7·61
(change to high)			+1·34		+3·34
High income			6·63	+4·32	10·95

*24-hour average weekday rates for LTS area residents of households with one employed resident in areas of medium rail and bus accessibility

3×3×3×3=81 categories. The average number of generations for each of these 81 categories was calculated and the number of households in each cell was estimated from the planning parameters for each district. The total number of generations was then obtained by multiplying the generation rates by the number of households in each category and summing for all categories. An example of the variation of trip generation rates with income and car ownership is shown in Table 2.6.

Trip attraction

2.21 The parameters used to estimate the number of 'attractions' or non-home ends of trips in each district were the amounts of employment in each district in six categories. A relationship was derived between employment and trip attraction for each district in 1962.

Person trips

Internal trips
(trips wholly within the LTS area)

2.22 Trip generation rates for 1962 are shown in Table 2.7 for different modes and purposes of travel. Car-owning households made an average of 6.06 trips per day whilst non-car-owning households made 2.47 compared with an average of 3·83 for all households.

2.23 Table 2.8 gives the total number of trips by trip purpose and mode of travel for the LTS area, and Table 2.9 gives similar data for the Central Area. Note the much higher proportion of British Rail and underground trips for the Central Area than for the LTS area as a whole and the lower proportion of car trips to the Central Area. More trips were by bus than any other mode for both the Central Area and the LTS area. Table 2.10 summarises the pattern of internal and external trips subdivided in each case between those trips with a destination (attraction) in the Central Area and the non-Central Area.

External trips
(trips between the LTS area and a point outside)

2.24 Because of the difficulty of obtaining accurate information on the movements of non-residents of the LTS area within the area, information on external trips, particularly for public transport, is not as complete or as accurate as the information on internal trips. The number of reported external trips leaving and entering the LTS area amounted to less than 10 per cent of all trips and on this basis any errors in the road traffic analysis were probably insignificant. However, many external trips by public transport were not reported, neither were non-home-based trips by any external traffic. Furthermore, although not strictly speaking external, the residents of hotels and institutions were not included in the home interview survey (partly because they were not thought to be significant car users) and they would be likely to make an appreciable number of public transport trips during the day. Although these omissions were probably not significant on the radial routes into London, they do form an appreciable element of the loadings on links in the Central Area and to alleviate the effects of the absence of these trips in Phase III of the LTS a specific estimate was made of the 1962 unreported trips which was then taken into account for the 1981 results.

Commercial vehicle and taxi trips

Commercial vehicle trips

2.25 The analysis of commercial vehicle trips is a more difficult task than that of person trips because there is no convenient basis for analysis corresponding to the household. The problem is further compounded by the extreme

Table 2.7 Rates of internal trip generation by main mode of travel and basic trip purpose—car-owning and non-car-owning households (Phase I)

	A—Car-owning households				*B—Non-car-owning households*				*C—All households*			
Main mode of travel	*Number of trips**	*Percentage*	*Trips per person*	*Trips per household*	*Number of trips**	*Percentage*	*Trips per person*	*Trips per household*	*Number of trips**	*Percentage*	*Trips per person*	*Trips per household*
Car driver	3,620,437	53·3	0·99	3·23	28,305	0·6	0·01	0·01	3,648,742	32·2	0·43	1·23
Car passenger	984,885	14·5	0·27	0·88	260,853	5·7	0·05	0·14	1,245,738	11·0	0·14	0·42
Cycle driver	134,947	2·0	0·04	0·12	335,409	7·4	0·07	0·18	470,356	4·2	0·05	0·16
Cycle passenger	11,387	0·2	0·00	0·01	35,173	0·8	0·01	0·02	46,560	0·4	0·01	0·02
Bus	1,198,049	17·6	0·33	1·07	2,692,249	59·4	0·55	1·47	3,890,298	34·3	0·46	1·32
Underground	419,902	6·2	0·12	0·37	658,012	14·5	0·13	0·36	1,077,914	9·5	0·13	0·36
British Rail	341,145	5·0	0·09	0·30	415,243	9·2	0·08	0·23	756,388	6·7	0·09	0·26
Taxi	34,252	0·5	0·01	0·03	35,491	0·8	0·01	0·02	69,743	0·6	0·01	0·02
Coach	29,924	0·4	0·01	0·03	25,860	0·6	0·01	0·01	55,784	0·5	0·00	0·02
Commercial vehicle passenger	23,327	0·3	0·01	0·02	47,425	1·0	0·01	0·03	70,752	0·6	0·01	0·02
Total	6,798,255	100·0	1·87	6·06	4,534,020	100·0	0·93	2·47	11,332,275	100·0	1·33	3·83
Basic trip purpose	*Number of trips**	*Percentage*	*Trips per person*	*Trips per household*	*Number of trips**	*Percentage*	*Trips per person*	*Trips per household*	*Number of trips**	*Percentage*	*Trips per person*	*Trips per household*
Work	2,631,152	38·7	0·72	2·34	2,746,028	60·6	0·57	1·50	5,377,180	47·5	0·63	1·82
Employer's business	173,828	2·6	0·05	0·15	50,860	1·1	0·01	0·03	224,688	2·0	0·03	0·08
Personal business	1,083,861	15·9	0·30	0·97	299,863	6·6	0·06	0·17	1,383,724	12·2	0·16	0·47
Entertainment	360,082	5·3	0·10	0·32	243,736	5·4	0·05	0·13	603,818	5·3	0·07	0·20
Sport	62,478	0·9	0·02	0·06	42,797	0·9	0·01	0·02	105,275	0·9	0·01	0·04
Social	494,997	7·3	0·14	0·44	317,393	7·0	0·06	0·17	812,390	7·2	0·10	0·27
Shopping—convenience	367,191	5·4	0·10	0·33	292,482	6·5	0·06	0·16	659,673	5·8	0·08	0·22
Shopping—hard goods	149,684	2·2	0·04	0·13	126,335	2·8	0·03	0·07	276,019	2·4	0·03	0·09
School	398,238	5·9	0·11	0·35	151,490	3·3	0·03	0·08	549,728	4·9	0·06	0·19
Miscellaneous	1,076,744	15·8	0·29	0·96	263,036	5·8	0·05	0·14	1,339,780	11·8	0·16	0·45
Total	6,798,255	100·0	1·87	6·05	4,534,020	100·0	0·93	2·47	11,332,275	100·0	1·33	3·83

*24-hour average weekday internal trips by LTS area residents of households

Table 2.8 LTS area internal basic trips* by trip purpose and main mode of travel (Phase I)

Main mode of travel	*Work*	*Other home-based*	*Non-home-based*	*Total*	*Percentage*
Car driver	1,239,147	1,583,603	825,992	3,648,742	32·2
Car passenger	317,652	791,463	136,623	1,245,738	11·0
Cycle driver	304,131	117,361	48,865	470,357	4·2
Cycle passenger	22,489	21,493	2,578	46,560	0·4
Bus	2,044,301	1,669,309	176,688	3,890,298	34·3
Underground	780,734	220,756	76,424	1,077,914	9·5
British Rail	618,746	117,950	19,692	756,388	6·7
Taxi	9,877	39,036	20,830	69,743	0·6
Coach	16,428	35,465	3,890	55,783	0·5
Commercial vehicle passenger	23,676	18,879	28,197	70,752	0·6
Total	5,377,181	4,615,315	1,339,779	11,332,275	100·0

*24-hour average weekday internal trips by LTS area residents of households

Table 2.9 Central area basic trips* by trip purpose and main mode of travel (Phase I)

Main mode of travel	*Work*	*Other home-based*	*Non-home-based*	*Total*	*Percentage*
Car driver	176,796	101,308	103,238	381,342	14·5
Car passenger	54,157	55,165	20,901	130,223	5·0
Cycle driver	43,250	8,557	13,690	65,497	2·5
Cycle passenger	4,240	1,008	375	5,623	0·2
Bus	577,753	157,139	62,683	797,575	30·3
Underground	538,366	95,509	49,019	682,894	25·9
British Rail	462,112	39,259	3,595	504,966	19·2
Taxi	7,682	24,386	17,608	49,676	1·8
Coach	714	2,991	832	4,537	0·2
Commercial vehicle passenger	3,001	2,956	4,712	10,669	0·4
Total	1,868,071	488,278	276,653	2,633,002	100·0

*24-hour average weekday internal trips by LTS area residents of households

range of use to which commercial vehicles are put, the difficulty of differentiating between private vehicles and light commercial vehicles and the existence of large numbers of very short trips by milk floats and the like. However it is possible to examine the pattern of movement by breaking it into trips within zones and trips between zones for light and heavy commercial vehicles. In 1962 a total of 191,000 commercial vehicles made 3·7 million trips per day of which retail delivery represented over 50 per cent of the total trips and about one quarter of the total commercial vehicle mileage in the LTS area. The breakdown of these trips for light vans, which are called light commercial vehicles, and for all other commercial vehicles over 30 cwt, which are called heavy, is shown in Table 2.11. Light commercial vehicles averaged about 22 miles per day and heavy commercial vehicles about 29 miles per day. Only about 8 per cent of the total number of commercial vehicles were heavy vehicles with three or more axles.

2.26 Figure 2.12 illustrates the density of internal commercial vehicle destinations by district and shows the highly concentrated distribution of trip ends towards the centre of London. The external commercial vehicle movements on the other hand have destinations which are more evenly spread and of a very much lighter intensity. Table 2.10 gives the total number of commercial vehicle trips within the LTS area on an average weekday. It will be seen that external trips are a very small proportion of the total number of commercial movements.

Taxi trips

2.27 The number of trips by taxi is very small compared with the number by other modes. The important point to note about taxi trips is their concentration on the Central Area: 66 per cent have at least one end in the Central Area and 55 per cent are wholly within the Central Area.

Table 2.10 Pattern of basic trips* (Phase I)

	Driver and other	*Bus and rail*	*Commercial vehicle*
Central area internal trips	648,000	1,985,000	3,702,000 (central and non-central internal)
Non-central area internal trips	4,960,000	3,739,000	
Central area external trips	55,000	495,000	74,000 (central and non-central external)
Non-central area external trips	603,000	249,000	
Through trips	20,000	3,000	9,000
Total	6,286,000	6,471,000	3,785,000

*24-hour average weekday trips
Coach trips—internal trips are included in 'driver and other',—external and through trips are included in 'bus and rail'

Table 2.11 Commercial vehicle trips* (Phase I)

	Light commercial		*Heavy commercial*		*Total*	
Type of trip	*Number*	*Percentage*	*Number*	*Percentage*	*Number*	*Percentage*
Inter-zonal	0·62	17	0·44	12	1·06	29
Intra-zonal	2·02	54	0·62	17	2·64	71
Total	2·64	71	1·06	29	3·70	100

*24-hour average weekday million internal trips

Figure 2.12 Density of internal commercial vehicle destinations by traffic district (Phase I)

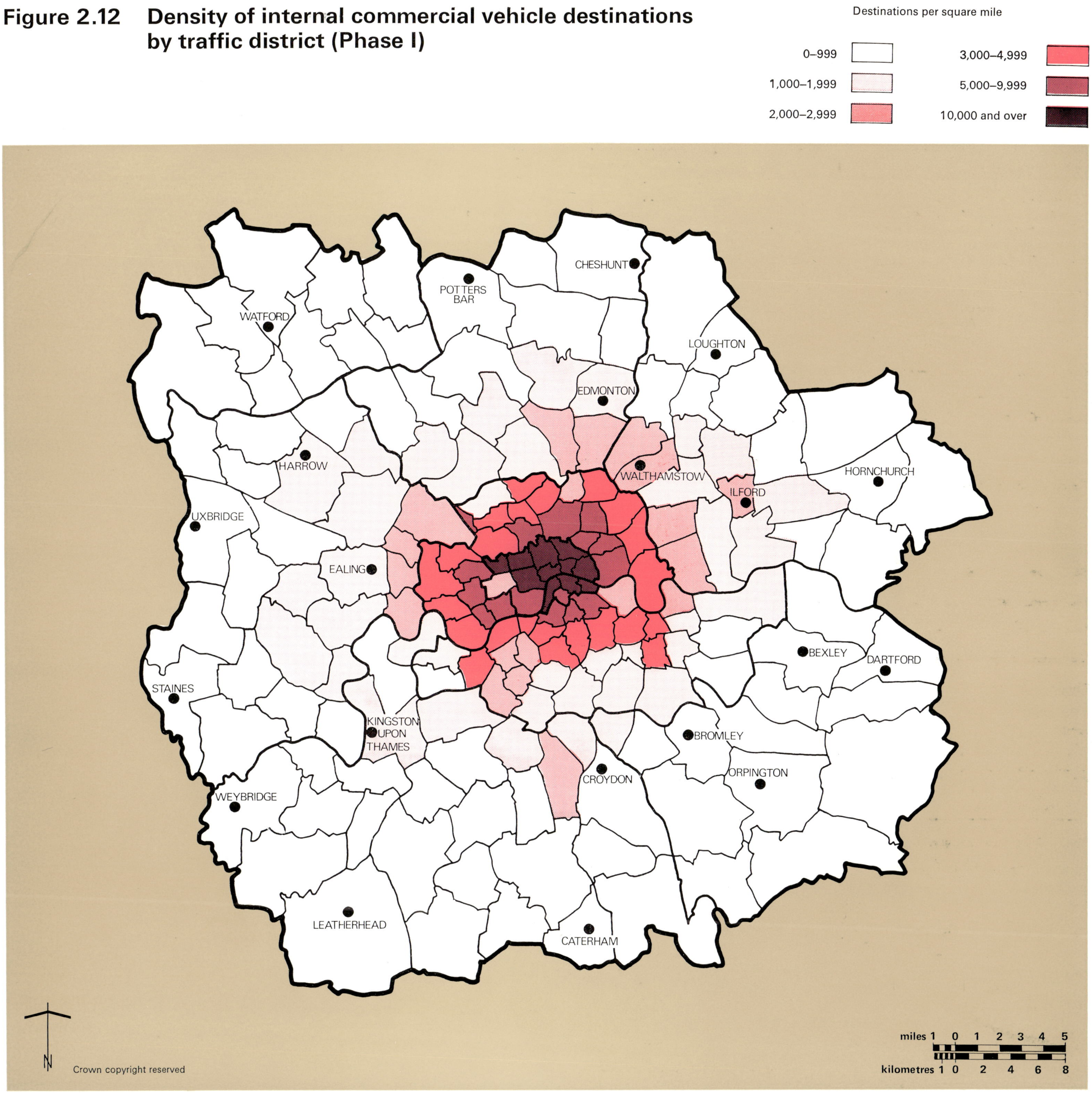

3

Chapter 3 Potential demand in 1981 (Phase II)

Introduction

3.1 In Phase II of the LTS the first estimates were made of the traffic in 1981. These were not forecasts of what would happen in that year but estimates of the demand for movement if population, employment and incomes changed in an assumed way between 1962 and 1981 and if people were able to and did behave in the same way as they had in 1962. Thus estimates of potential demand were made and for the first time the size of the transport problem that London faced was quantified.

3.2 The procedure adopted in Phase II followed the lines indicated in Figure 2.1. A division between public and private transport was made for internal trips but the public transport analysis was carried no further.

3.3. The first stage of the forecasting technique was to derive relationships using the basic data collected during Phase I. Using these relationships future numbers of trips were estimated on the basis of new planning data for the forecast year. At this stage indices of public transport accessibility were used in the analysis to divide the trips between public and private transport. The pattern of travel was calculated using a trip distribution model calibrated from 1962 conditions and using speeds for the primary roads that might reasonably be expected in uncongested conditions and speeds on secondary roads similar to those observed in 1962. Finally the distributed trips were assigned to a network on the assumption that they would follow the shortest time path between origin and destination. This has been referred to as a desire assignment, being that pattern of movement which might be expected to occur should there be no impediment to movement in terms of congestion beyond the 1962 level.

3.4 The forecast planning data prepared for and used in Phase II were also used in Phase III. They have subsequently been revised by the GLC and used in a test which is described in Chapter 5. Further revisions in line with estimates for the GLDP are being used at present for the latest series of tests.

Forecasts of planning data

Population

3.5 Table 3.1 summarises the population forecast for the LTS area which was used in Phases II and III compared with that for the rest of the country. Small increases in population were forecast for the LTS area, predominantly towards the boundary of the area. Large increases were anticipated just outside the boundary. Table 3.2 shows how these population changes are reflected in the distribution of households and the average number of persons per household at the sector level.

Employment

3.6 The major changes forecast in the distribution of employment by industrial

Table 3.1 Population forecasts—LTS area, London Metropolitan region and Great Britain (Phases II and III)

Planning area	*1981 population*		*Percentage change 1961–1981*
	Number	*Percentage*	
LTS area			
Private households	8,815,000	14·8	+3·6
Other	326,000	0·6	+1·6
Total LTS area	9,141,000	15·4	+3·6
Remainder of London Metropolitan region	5,368,000	9·0	+47·5
Total London Metropolitan region	14,509,000	24·4	+16·4
Remainder of South-East England	6,858,000	11·5	+31·9
Total South-East England	21,367,000	35·9	+20·9
Remainder of Great Britain	38,138,000	64·1	+13·4
Total Great Britain	59,505,000	100·0	+16·0

Figure 3.1 Employment density by traffic district (Phases II and III)

Employment density per square mile

0–199	5,000–19,999
200–999	20,000–99,999
1,000–4,999	100,000 and over

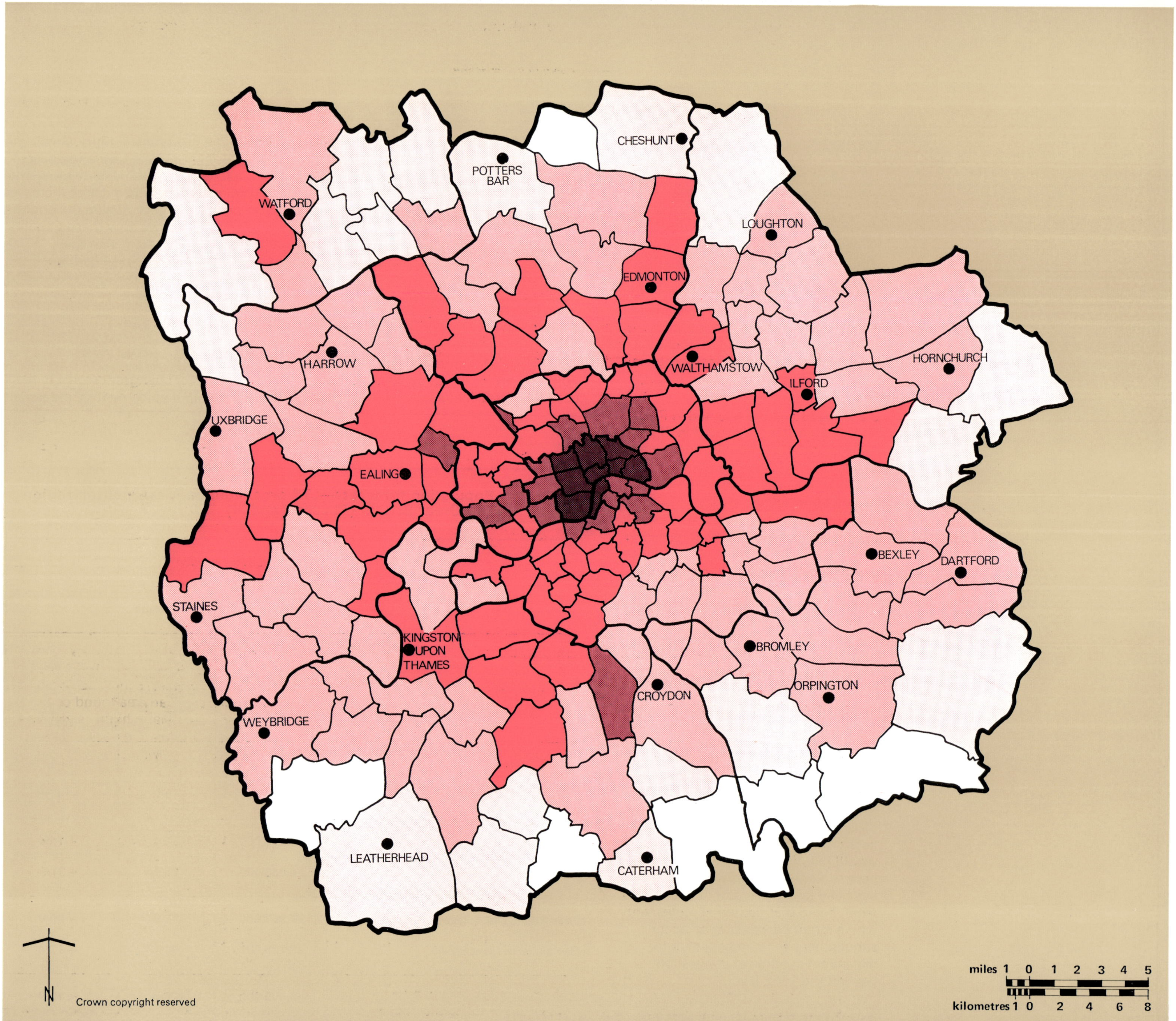

Table 3.2 Distribution of households and population in households by sector (Phases I, II and III)

Traffic sector	Average number of persons per household 1961	1981 population in households			1981 households			Average number of persons per household 1981
		Number	Percentage	Percentage change 1961–81	Number	Percentage	Percentage change 1961–81	
0	3·00	1,333,000	15·1	−1·11	466,000	14·9	+3·85	2·86
1	2·93	976,000	11·1	+3·71	336,000	10·8	+4·72	2·90
2	3·18	263,000	3·0	+19·17	83,000	2·7	+19·72	3·17
3	3·06	1,187,000	13·5	−3·44	414,000	13·2	+3·02	2·87
4	2·55	1,393,000	15·8	+1·66	536,000	17·1	−0·18	2·60
5	2·22	199,000	2·2	+4·30	90,000	2·9	+4·65	2·21
6	2·67	31,000	0·4	+2·85	12,000	0·4	+6·38	2·58
7	2·86	1,562,000	17·7	+9·07	542,000	17·3	+8·38	2·88
8	3·05	618,000	7·0	+6·31	213,000	6·8	+11·72	2·90
9	2·96	1,253,000	14·2	+7·85	434,000	13·9	+10·37	2·89
Subtotal central area	2·27	230,000	2·6	+4·10	102,000	3·3	+4·96	2·25
Total LTS area	2·87	8,815,000	100·0	+3·64	3,126,000	100·0	+5·64	2·82

grouping are shown in Table 3.3. Manufacturing and wholesaling employment were not expected to alter very much but a large increase was forecast in retail employment and in other employment, which includes the majority of the service industry sector and office employment. Table 3.3 also shows the data for the Central Area as distinct from the LTS area as a whole.

3.7 A comparison of Figures 3.1 and 2.4 shows very similar distributions of employment density by district with only limited changes in certain districts. The most important aspects of these changes were the large increases in districts to the north-west and south of central London whilst in the east of the old LCC area there were some decreases in employment. Finally Table 3.4 illustrates for both 1962 and 1981 the number of employed persons per household by sector. The changes in individual sectors and for the LTS area as a whole were very small.

Income

3.8 The National Economic Development Council targets were used as a basis for the income forecasts. These involved a 4 per cent annual increase in the GNP and a 2·8 per cent annual increase in consumer buying power. The equivalent overall mean household income forecast was 3·25 per cent per annum or an increase of 85 per cent in real income between 1962 and 1981. Table 3·5 shows the swing from the lower income groups to the higher income groups and the average income in each district is shown in Figure 3·2. A comparison of Figures 3.2 and 2.7 illustrates the extent of these changes, particularly in the north, the west and the south-west. There would also be considerable changes within the high density urban area round central London, a fact that is highly significant because of the relationship between income and car ownership.

Car ownership

3.9 The basic relationship between car ownership and household income used in the LTS is shown in Figure 3.3. The future split between car-owning and non-car-owning households derived from this relationship is shown in Table 3.6. It was estimated that 68 per cent of all households in the area would have the use of one or more cars which compares with 38 per cent in 1962. The distribution of car-owning households

Table 3.3 Distribution of employment by industrial group (Phases II and III)

Industrial group	*Number of employees 1981*	*Percentage change 1962–1981*
LTS area:		
Manufacturing	1,504,000	−9·8
Retailing	596,000	+33·3
Wholesaling	252,000	+7·7
Other	2,965,000	+22·3
Total	5,317,000	+11·4
Central area:		
Manufacturing	323,000	−5·3
Retailing	104,000	+11·8
Wholesaling	123,000	−2·4
Other	1,086,000	+22·9
Total	1,636,000	+13·3

Table 3.4 Employed persons per household by sector of residence (Phases I, II and III)

Traffic sector of residence	*All households*	
	1962	*1981*
0	1·49	1·49
1	1·47	1·45
2	1·56	1·53
3	1·50	1·48
4	1·40	1·41
5	1·20	1·20
6	1·50	1·42
7	1·44	1·46
8	1·38	1·37
9	1·37	1·36
Total LTS area	1·43	1·43

Table 3.5 Distribution of LTS area households by household income class (Phases II and III)

Income class	*Range of total annual household income*	*1981*		*Percentage change 1962–81*
		Number of households	*Percentage*	
1	under £500	146,000	4·7	−65
2	£500–£999	475,000	15·2	−60
3	£1,000–£1,499	713,000	22·8	−13
4	£1,500–£1,999	600,000	19·2	+90
5	£2,000–£2,999	756,000	24·2	+380
6	£3,000 and over	436,000	13·9	+450
		3,126,000	100·0	+5·64

Table 3.6 Percentage of car-owning and non-car-owning households by sector (Phases I, II and III)

Traffic sector	*1962*		*1981*	
	Car-owning	*Non-car-owning*	*Car-owning*	*Non-car-owning*
0	44·1	55·9	77·6	22·4
1	43·7	56·3	76·6	23·4
2	48·7	51·3	82·3	17·7
3	38·4	61·6	71·1	28·9
4 & 5	26·9	73·1	51·3	48·7
6 & 7	29·2	70·8	54·3	45·7
8	46·4	53·6	80·9	19·1
9	48·5	51·5	81·5	18·5
Total LTS area	37·9	62·1	68·1	31·9

Figure 3.2 Mean household income by traffic district (Phases II and III)

Mean household income (£)

1,000–1,499	
1,500–1,999	
2,000–2,499	
2,500–2,999	
3,000 and over	

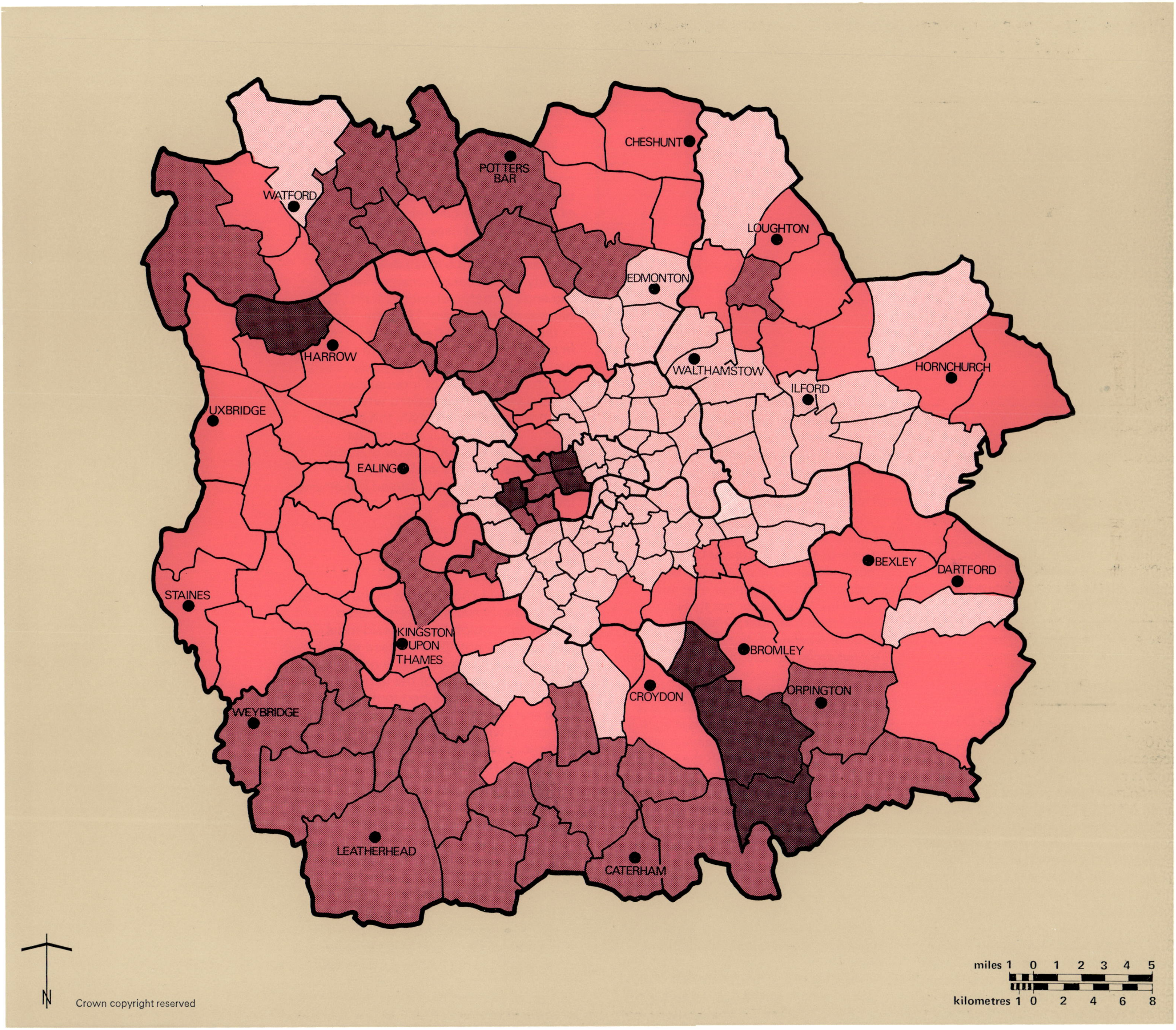

Figure 3.3 A car ownership and household income relationship (Phases II and III)

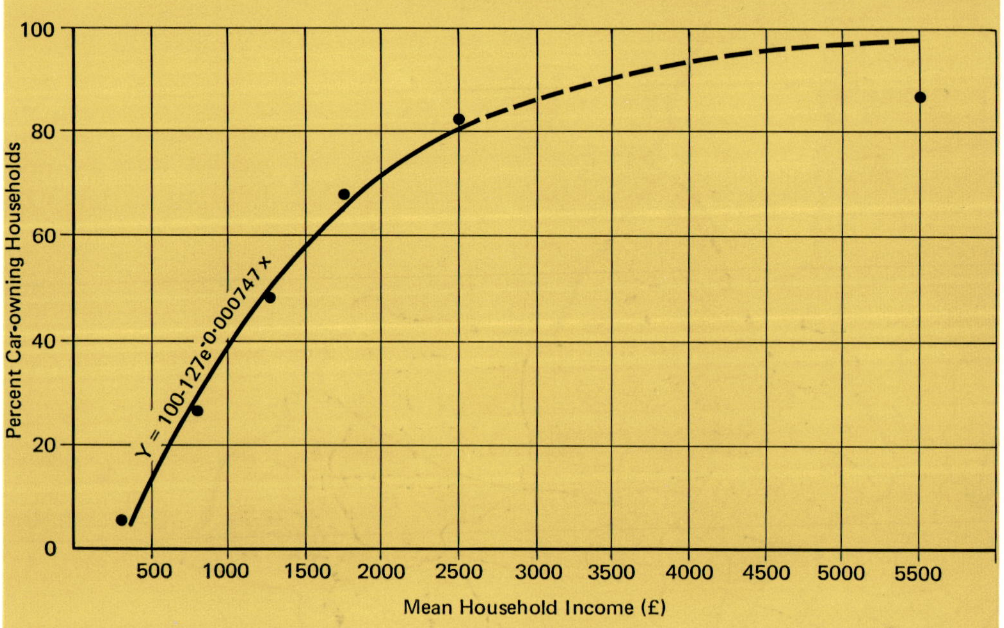

is shown in Figure 3.4. The most rapid increase in car ownership is expected in areas with lower than average per capita income and higher than average density of population.

Commercial vehicle, taxi and external trips

3.10 There is a number of reasons why forecasting commercial, taxi and external trips is not so precise as forecasting person movement internal to the LTS area. A 1981 matrix or pattern of road movements was therefore produced separately for each of these three categories and they were then added together. The summed matrix was used in Phase II and has since been used in Phase III and in subsequent work by the GLC.

3.11 There is some evidence which suggests that commercial vehicle trips are at present decreasing in number in certain parts of London. This could well be as a result of an apparent trend towards the use of fewer larger commercial vehicles and the consequent reduction in the number of small vehicles. Whether or not this represents more than a temporary halt, for technological reasons, in the rapid growth of commercial vehicle trips is not yet clear.

3.12 The impact of growth outside London is difficult to assess but it is unlikely that any development such as the Channel Tunnel will have a significant impact as compared with the volume of London's own traffic. However the work at present being undertaken for the study of the South-East region will provide information which will allow the existing external trip forecasts to be checked.

Commercial vehicle trips

3.13 Commercial vehicle trips were forecast using regression equations relating trip generation to vehicle utilisation and to zonal characteristics. For external trips growth factors were estimated in relation to alterations in employment levels both inside and outside the LTS area and these were used to adjust the flows measured in 1962.

3.14 The results of these forecasts are shown in Table 3.7. Even ignoring the very short intra-zonal trips the increase in light commercial vehicle trips associated with residences and other consumer-orientated land uses is the biggest of all. The next biggest is that of heavy commercial vehicles associated with similar land uses. A total increase in inter-zonal trips of 73 per cent was forecast. Figure 3.5 illustrates the density of internal commercial vehicle trip ends by traffic district in 1981. The biggest increases over 1962 are generally in the outer areas reflecting the increase in population in these areas. However, these changes have not seriously affected the density of internal commercial vehicle destinations which are still orientated very heavily towards the Central Area. External commercial vehicle trips, although fewer than the internal commercial vehicle trips, are very much more evenly spread out among internal destinations.

Taxi trips

3.15 Phase I showed that in the Central Area home-based trips were related closely to the high income residential areas and the non-home-based to hotels, stations, clubs, airline terminals and offices. Because of the large number of taxi trips which are tied to the Central Area and the very small number of these compared with internal vehicle trips as a whole, it was decided to keep the taxi trips at the 1962 level.

External private trips

3.16 In calculating external trips the technique adopted was to calculate growth factors for trips by residents and non-residents related to the anticipated changes in population and employment at the ends of the trips. After applying these factors to the 1962 trip ends they were distributed utilising specially calibrated distribution models to produce a matrix of external trips.

Figure 3.4 Car ownership by traffic district (Phases II and III)

Percentage of car owning household

40–59·9

60–79·9

80 and over

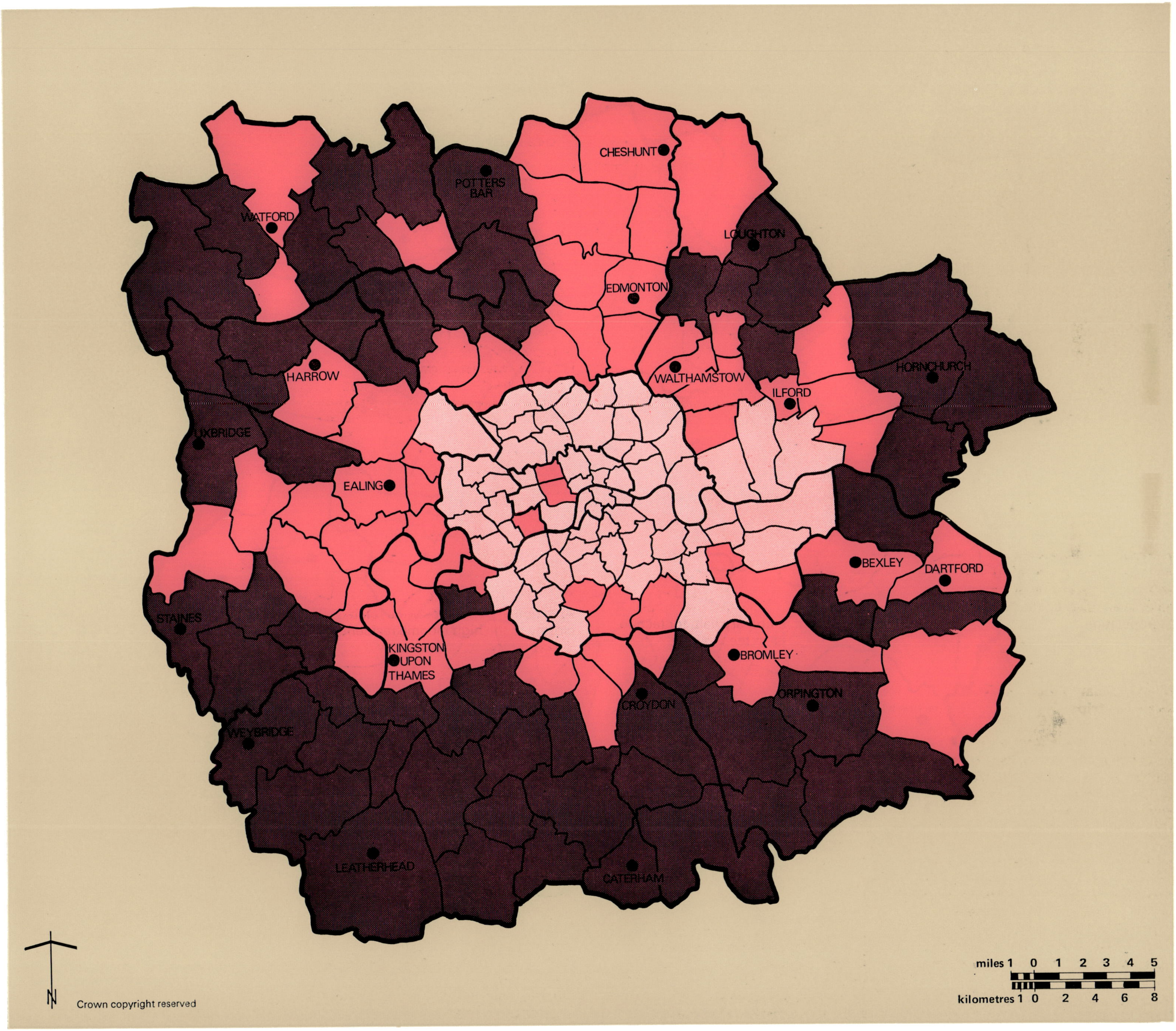

Figure 3.5 Density of internal commercial vehicle destinations by traffic district (Phases II and III)

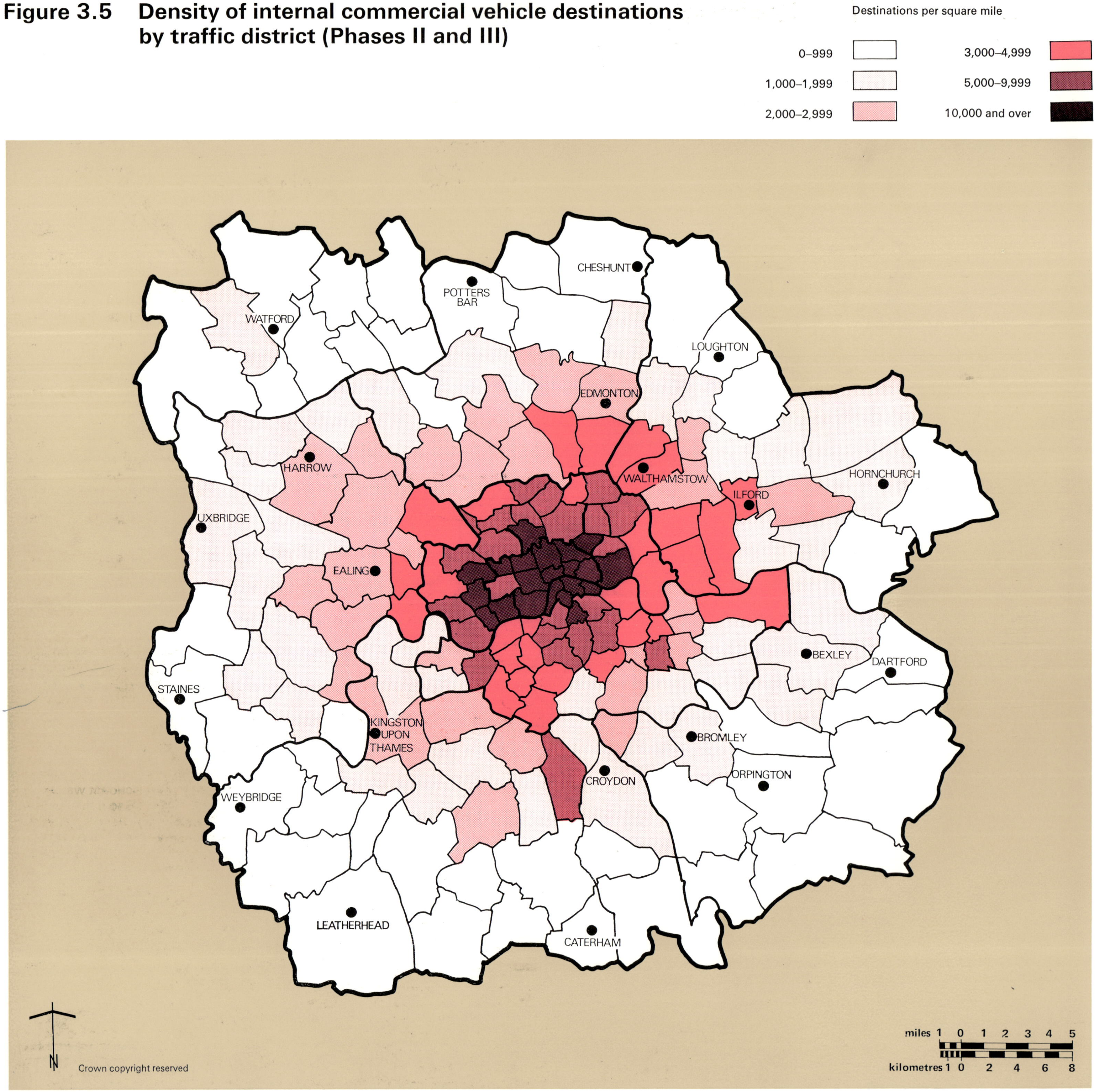

Table 3.7 Internal inter-zonal trips* by light and heavy commercial vehicles by land use group (Phases II and III)

Land use group	*Number of trips per day 1981*	*Percentage increase 1962–1981*
***A* Light commercials vehicles to**		
Residences and other consumer orientated land uses	1,036,742	124
Industry and other non-consumer orientated land uses	168,930	20
Subtotal light commercials	1,205,672	100
***B* Heavy commercials vehicles to**		
Residences and other consumer orientated land uses	391,135	50
Industry and other non-consumer orientated land uses	169,721	10
Subtotal heavy commercials	560,856	35
Total	1,766,528	73

*24-hour average weekday inter-zonal trips by commercial vehicles based in the LTS area

Networks developed

3.17 In Phase II, two road networks were developed and these are referred to as the 1981A network, which is illustrated in Figure 3.6, and the 1981B network. The latter represented a minimum investment network drawn up on the basis of known programmes of work in the then LCC and the Ministry of Transport. The 1981A network represented a more complete network and arose from work done following the publication of the Abercrombie Plan and consideration of the geographic and environmental constraints on the location of new roads in London. The final results from Phase II, following the distribution and assignment, are shown in Figure 3.7 in the traffic volumes loaded on to the 1981A road network. These loadings, calculated using speeds on the primary network assuming no congestion, reflect a desire assignment based on potential demand, the way people could be expected to behave in a situation constrained to no greater degree than existed in 1962. They indicated both the massive scale of the potential demand for movement and the need to develop a balanced programme of investment as well as certain restraint policies to control the demand.

The need for further study

3.18 Although the results of Phase II provided a basis for action it was necessary to extend the LTS to a third phase. The purpose of Phase III was to improve and extend Phase II so as to provide:

a. A forecast of the pattern of public transport demand.

b. Tests of alternative combinations of road and public transport networks.

c. A realistic estimate of volumes of travel by car in which the demand was in balance with the capacity of the networks tested.

d. An economic evaluation of the networks tested.

3.19 The techniques developed to meet these further needs and the results of their use are described in the next chapter.

Figure 3.6 1981A network (Phase II)

Figure 3.7 1981 traffic assigned to the 1981A network (Phase II)

Traffic scale

Thousands of vehicles *(per 24-hour average weekday)*

500 400 300 200

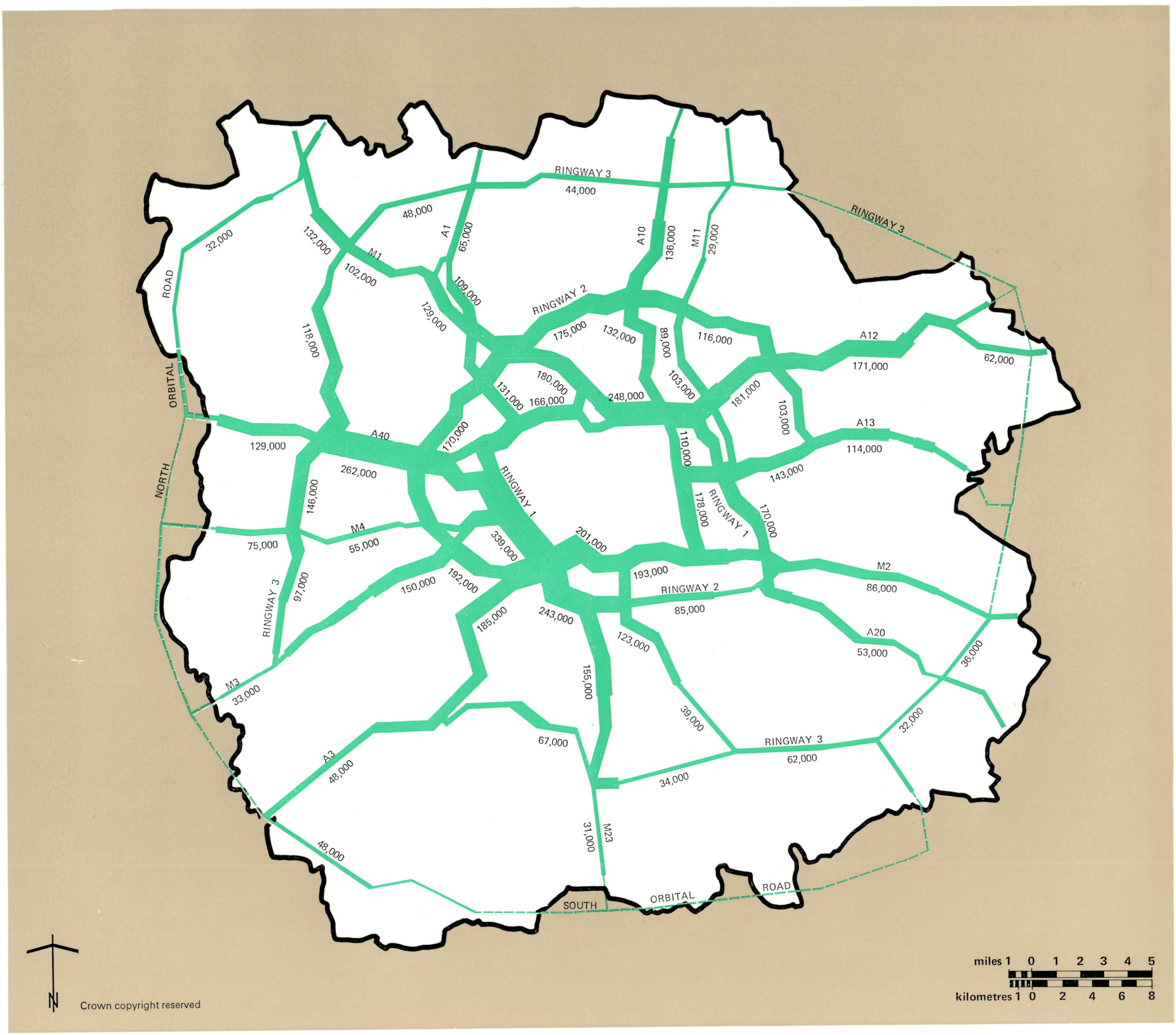

4

Chapter 4 Constrained demand in 1981 (Phase III)

The objectives

4.1 The broad objectives of Phase III were to study further the basic policy issues raised by Phases I and II and to examine some possible alternative transport networks. To do so it was necessary to produce a model that related trip generation to transport capacity and in which road traffic assignments were sensitive to congestion. An economic analysis of the costs and benefits of transport improvement was to be included, which meant that a range of road and public transport plans was required to enable comparisons to be made. There was also the need to develop procedures to forecast patterns of movement by public transport which had not been done in Phase II.

4.2 Tasks of this kind had never before been attempted on such a scale. Phase III consequently had a high research content and the extent of the work possible within the time available was largely determined by the outcome of this research. Partly for this reason and partly because of difficulties in adequately defining alternative land use plans it was decided to test only those transport plans related to the most likely land use plan. In the event, Phase III, while not solving all the technical problems it faced, took major steps towards solving each of the problems tackled and has provided a much improved basis for the assessment of realistic travel demands.

The procedure

4.3 A simplified flow diagram for the Phase III model is illustrated in Figure 4.1 and its general relationship with the whole process can be seen from Figure 2.1. The technical details of certain parts of the model are discussed in Appendix A. The model as it was developed has seven main elements:

a. Trip generation
b. Trip distribution
c. Modal choice
d. Traffic assignment
e. Speed reduction
f. Restraint procedure
g. Economic and operational analyses.

The analyses were made for the three trip purposes (work, other home-based, non-home-based) and two main modes (private, public) with commercial vehicle and taxi trips from Phase II and were made on a 24-hour basis except for public transport assignments which were made for the peak hour. The peak hour matrix was obtained by factoring 24-hour flows.

4.4 The trip generation procedure estimated the total travel demand by car-owning and non-car-owning households separately. Generations were grouped into categories:

Cars owned

0, 1 and 2 plus

Household income

Low (up to £1,000 per annum)
Medium (£1,000–£2,000 per annum)
High (over £2,000 per annum)

Employed residents

0, 1 and 2 plus

Net residential density

0–35, 35–65 and over 65 persons per acre.

The analysis was further divided on a geographic basis by Central Area and non-Central Area. Attractions were based on Phase II equations but additional factoring was necessary to divide them between car-owning and non-car-owning households.

4.5 The car-owning household travel demand was considered separately in the model from that of non-car-owning households since modal choice is most relevant to the former. For car-owning households total travel was distributed using a gravity model similar to that used in Phase II. Travel time diversion curves were used to separate public and private transport trips. The private transport trips were then assigned to shortest time routes in the road network. The resulting loadings were likely to exceed the capacity of some of the links as in Phase II. This led to the introduction into the model of procedures designed to represent the effects of discouraging or preventing some potential road users from using their cars for particular trips. The trips were either shortened, transferred to public transport, or eliminated.

4.6 The first step was successively to reduce assumed speeds on the motorways until the overloads on the motorways and the secondary road system were approximately equal. At each stage trip distribution and modal choice calculations were carried out which resulted in the shortening of trip lengths and the transfer of some trips to public transport as road speeds decreased. Finally, a linear programming procedure was introduced which represented regulatory measures to control the number of trips that could be made to specific destinations in such a way that traffic on the motorways was approximately equated with their capacity. This is referred to as the restraint procedure.

4.7 Those private trips which now could not take place on the roads were assumed to behave as if from non-car-owning households, some using public transport and some not taking place at all. The latter resulted from the fact that non-car-owning households have lower trip generation rates than car-owning households.

Figure 4.1 The model procedure (Phase III)

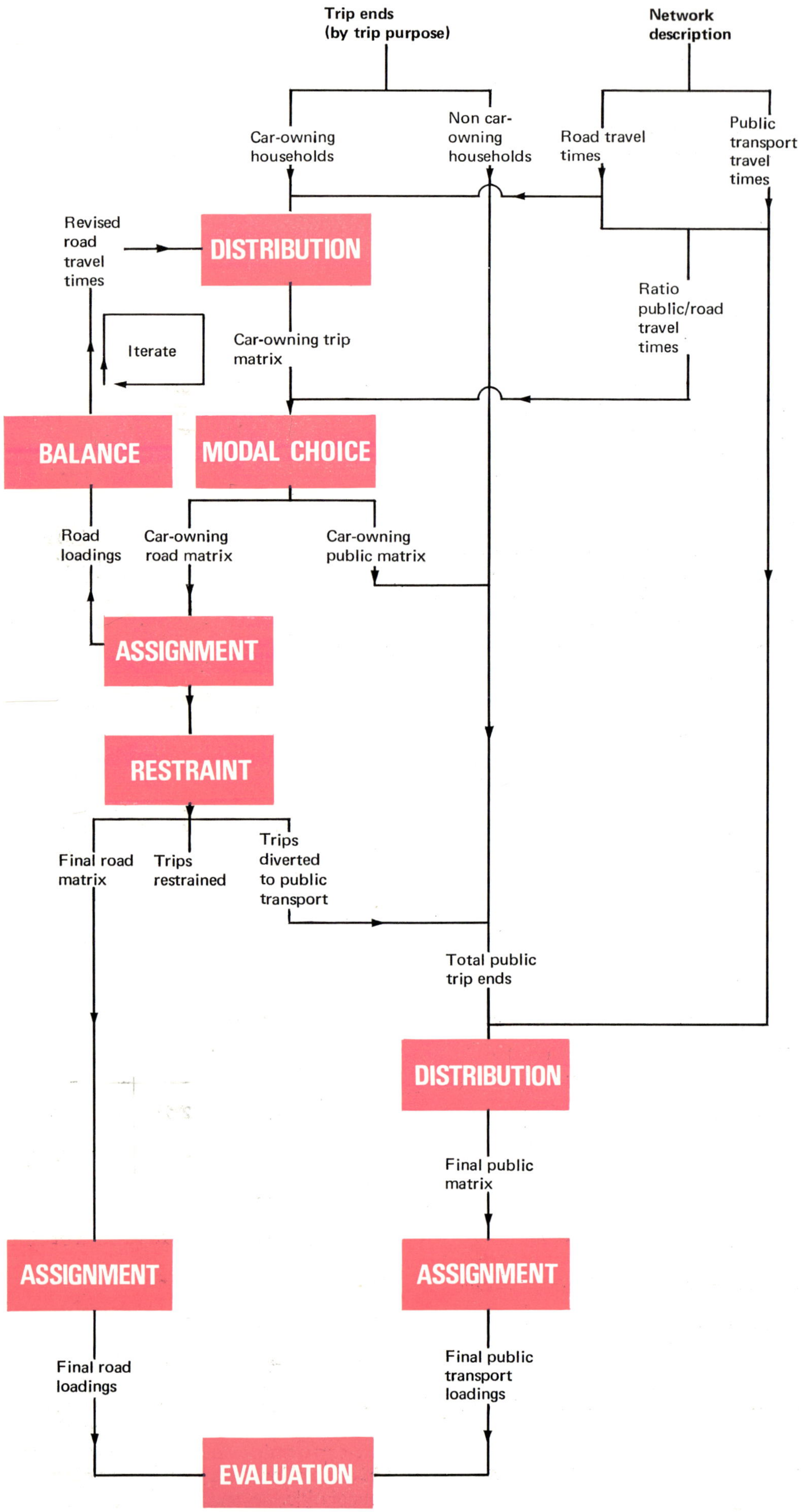

4.8 The public transport trips from car-owning households estimated during the modal choice calculations and the trips transferred to public transport by the restraint procedure were re-expressed as trip ends. These were added to the non-car-owning trip ends to give the total number of public transport trip ends. They were then distributed by a separately calibrated gravity model and assigned to the public transport networks. The public transport networks were defined in this phase in more detail than the road networks, using traffic zones rather than traffic districts. A special set of procedures, called Transitnet, was developed for analysis of public transport networks, taking account of the fact that public transport involves a system of regular services along fixed routes. Travel times, calculated during the Transitnet analysis, were used in calculating the modal choice of trips from car-owning households and as the basis of the distribution of public transport trip ends. Assignment of public transport trips used minimum time routes also derived in Transitnet. These procedures were not available for use in the early runs of the model and a simplified representation of public transport networks, based on techniques developed by the US Bureau of Public Roads (BPR), was used both to calibrate the modal choice diversion curves using 1962 data and in applying these curves to the 1981 networks.

4.9 For the reasons indicated previously in paragraph 2.24 the 1962 surveys had not included all public transport travellers. The objectives of Phase III necessitated estimates to be made of the number and pattern of unreported trips in 1962 by inspecting public transport travel records. An equivalent figure for 1981 was then estimated. Where overall results are quoted later in this chapter they include these unreported trips for Phases I, II and III.

4.10 The traffic model was used to analyse the operation of the road and public transport networks proposed for the future. The results of this traffic analysis were then used with economic data to compare costs and benefits in the design year.

Transportation plans developed

Road networks

4.11 Over many years, as the understanding of traffic patterns has developed, many physical feasibility studies have been made of motorway routes in London. This work was the basis of the development of the road network concept on which Phase II was based. Phase III road networks further refined these concepts with the knowledge of traffic patterns available from Phase II. In addition a wide range of theoretical networks was also developed based on the examination of traffic flows across imaginary grids. These are discussed in Appendix B. Eventually three networks were selected for evaluation. These three road networks were not intended specifically to represent official policy or plans but were possible alternatives to be compared in terms of their costs and ability to meet potential demand. The networks adopted for testing were:

a. *Plan 1*
representing one way of investing the minimum highway monies likely to be available by 1981.

b. *Plan 3*
a system of radial and circumferential routes.

c. *Plan 9*
a more extensive system likely to be able to cope with the potential demand in 1981.

Plans 1 and 3 are shown in Figures 4.2 and 4.3 and plan 9 in Appendix B.

4.12 While the mileage of motorways varied considerably between plans a uniformly high standard of geometric and environmental design was adopted for the purposes of the economic analysis. Thus, for example, the east side of Ringway 1 was costed and assumed to have a capacity of dual 4 lanes whereas in fact sections have been completed and are under construction with a capacity of dual 3 lanes. In addition the amount of land assumed to be taken by the motorways was greater than is the present practice, as a means of taking some account of environmental costs. For these reasons the estimates of capital costs of the plans are high compared with roads of this type that have been built in the past.

4.13 Plan 1 consisted of Ringway 3 around London linking the national motorways and, in addition, those motorways likely to be constructed by the early 1970's as a result of the programmes of work already committed at the time the plans were drawn up. This plan therefore does not result in a coherent system of motorways within Ringway 3 but in isolated motorways linked through the present system of roads.

4.14 Plan 3 formed a comprehensive system of motorways linking with the national motorways. It was similar in concept to the 1981A network of Phase II and consisted of Ringways 1, 2 and 3 and several radial motorways. It differed from the 1981A network in many details but the main differences were that a link within Ringway 1 in the east was omitted, the M.11 was re-aligned by the Roding Valley route rather than up the Lea Valley, Ringway 2 was re-aligned in the south-west, a Thames crossing was added in the east and Ringway 3 was re-aligned in the north-west and south-west.

4.15 In practice it will be appreciated that the scope available for construction of major roads within the Greater London area is severely limited. Constraints exist in the form of public open space, buildings of historic and architectural merit and topographical features. It is possible, however, to minimise community disruption by siting motorways parallel to railway lines and Figure 4.4 illustrates the extent to which the road and rail systems are adjacent in plan 3. Table 4.1 gives details of the actual land uses taken by the motorways in plan 3.

4.16 The capacity assumptions used in Phase III are shown in Table 4.2, corresponding to definitions of design and limiting capacity which are given in paragraph 4.49. Table 4.3 shows the extent of the mileage of motorways by number of lanes in plans 1 and 3. Plan 3 provides a considerable increase in capacity over plan 1. Note that the mileages differ slightly from those for the costed network in Tables 4.7, B.2 and C.1.

Secondary roads

4.17 For the purpose of the Phase III analysis it was necessary to consider the extent to which the secondary network

Table 4.1 Proportions of land uses plan 3 (Phase III)

Land use	*Ringway 1*	*Inner radial*	*Ringway 2*	*Outer radial*	*Ringway 3*	*Total*
A Residential	50%	40%	28%	18%	2%	19%
B Commercial	10%	10%	7%	1%	—	4%
C Industrial	4%	1%	1%	1%	—	1%
D Public	3%	21%	13%	13%	6%	10%
E Transportation	31%	25%	33%	20%	5%	17%
F Agricultural	—	1%	2%	38%	80%	42%
G Others	—	1%	2%	5%	6%	4%
H Vacant	2%	1%	14%	4%	1%	3%
Total	100%	100%	100%	100%	100%	100%

Figure 4.2 Road plan 1 (Phase III)

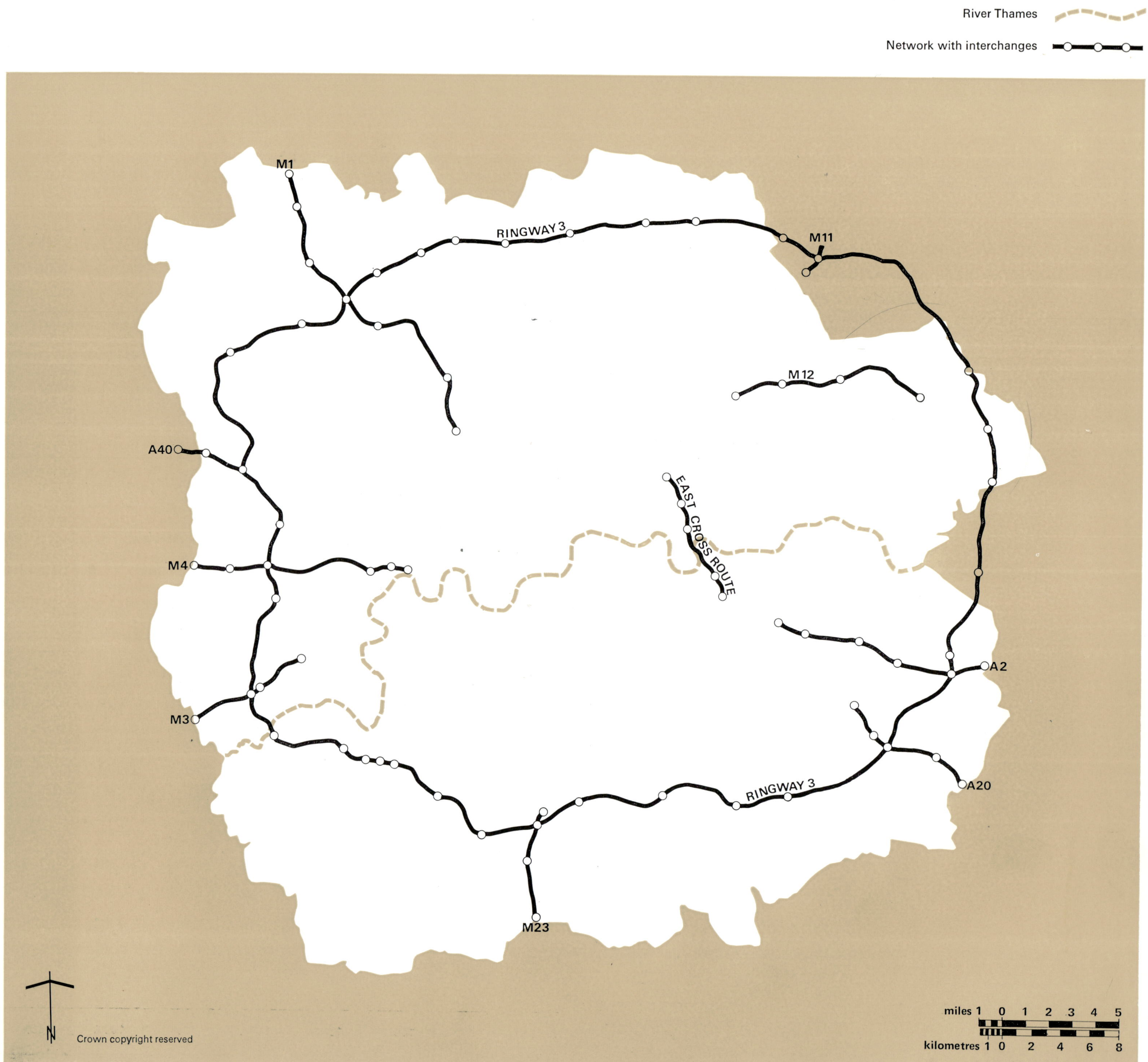

Figure 4.3 Road plan 3 (Phase III)

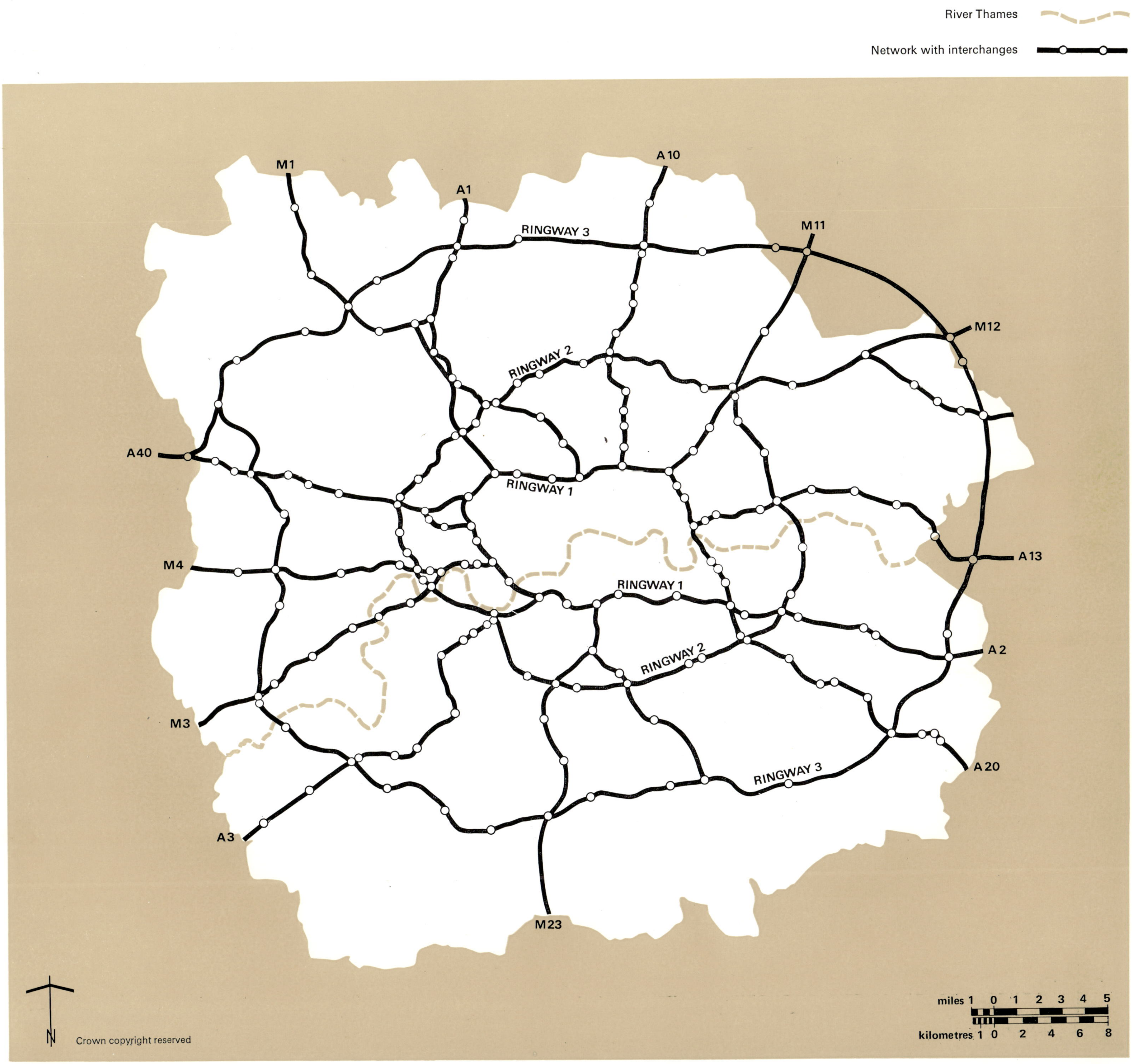

Figure 4.4 Motorway alignments adjacent to railway lines in plan 3 (Phase III)

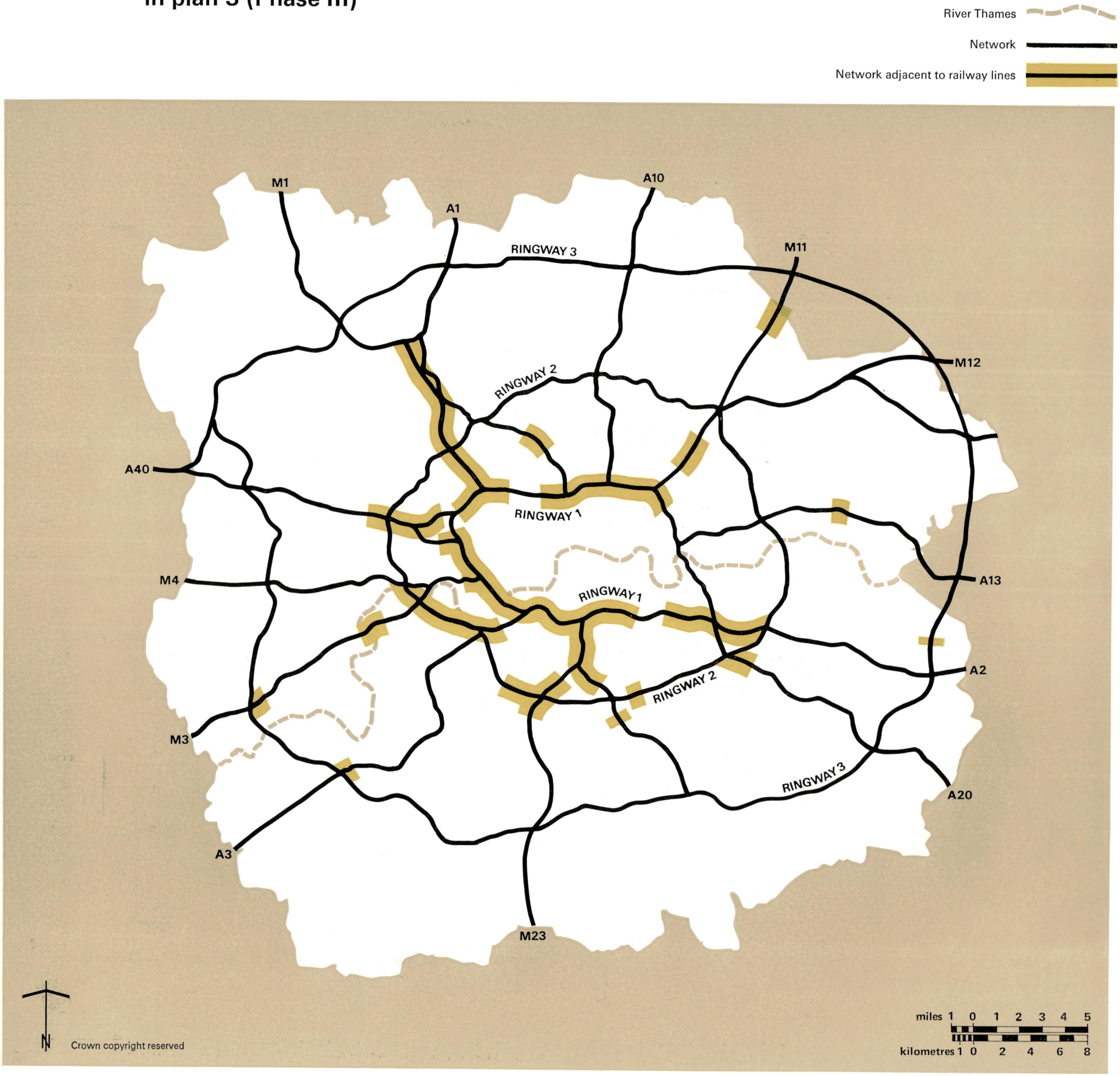

Table 4.2 24-hour design and limiting volumes of motorways (Phase III)

Number of lanes	Vehicles per day	
	Design	*Limiting*
2×2	66,000	75,000
2×3	100,000	112,000
2×4	133,000	150,000

would complement the primary network. Enough detail was needed to represent access to the primary system and to provide for traffic that would not use the primary system.

4.18 In order to determine the capacity of the secondary network the LTS area was divided into four smaller areas or jurisdictions. These were the Central Area, the remainder of the former LCC area, the remainder of the LTS area north of the Thames and the remainder of the LTS area south of the Thames. The capacities of the secondary road system were calculated for these jurisdictions based on observations of vehicle miles in the base year and assumed increases in capacity up to 1981. They are summarised in Table 4.4. Whilst these figures can be used to calculate the balance between the primary and secondary road networks as a whole it is still necessary to examine the detailed results to ascertain whether specific links of the secondary network are coping with the forecast demand.

Motorway interchanges

4.19 The provision of interchanges is an important consideration in designing motorway networks for a number of reasons:

a. They will probably be fairly large and have a considerable impact upon the environment in which they are built.

b. The presence of the public transport networks, frequently on viaducts or embankment, whilst providing ideal routes for the motorways to follow to minimise impact on the general environment, can cause severe problems for interchange design.

c. Interchanges between the motorways and the secondary roads are also crucial, in that they have a direct effect on motorway loadings. The greater the frequency at which they are provided the less critical their effect on the secondary roads themselves and the smaller the size of the junction can be. However they cannot be provided too close together or it will be impossible to operate the motorway networks themselves efficiently.

Table 4.3 Mileage of motorways (Phase III)

	Plan 1				*Plan 3*			
Number of lanes	2×2	2×3	2×4	*Total*	2×2	2×3	2×4	*Total*
Ringway 1	—	—	6	6	—	—	30	30
Ringway 2	—	—	—	—	—	7	42	49
Ringway 3	—	67	33	100	—	88	12	100
Inner radials (Ringway 1 to Ringway 2)	—	—	—	—	—	18	26	44
Outer radials (Ringway 2 to Ringway 3)	—	24	12	36	4	61	49	114
Total	—	91	51	142	4	174	159	337

Table 4.4 Capacities of secondary roads (Phase III)

Jurisdiction	*Area (square miles)*	*1962 Observed flow**	*1981 Estimated capacity of secondary network**	
			Plan 1	*Plan 3†*
Central area	10·4	1·958	2·350	2·350
Other LCC	105·9	5·863	6·964	6·616
Outer North	481·1	10·898	13·998	11·437
Outer South	343·6	5·004	7·221	6·681
Total	941·0	23·723	30·533	27·084

*24-hour average weekday millions of vehicle miles
†Some plan I secondary roads upgraded to primary in plan 3

Figure 4.5 Area of interchanges by location (Phase III)

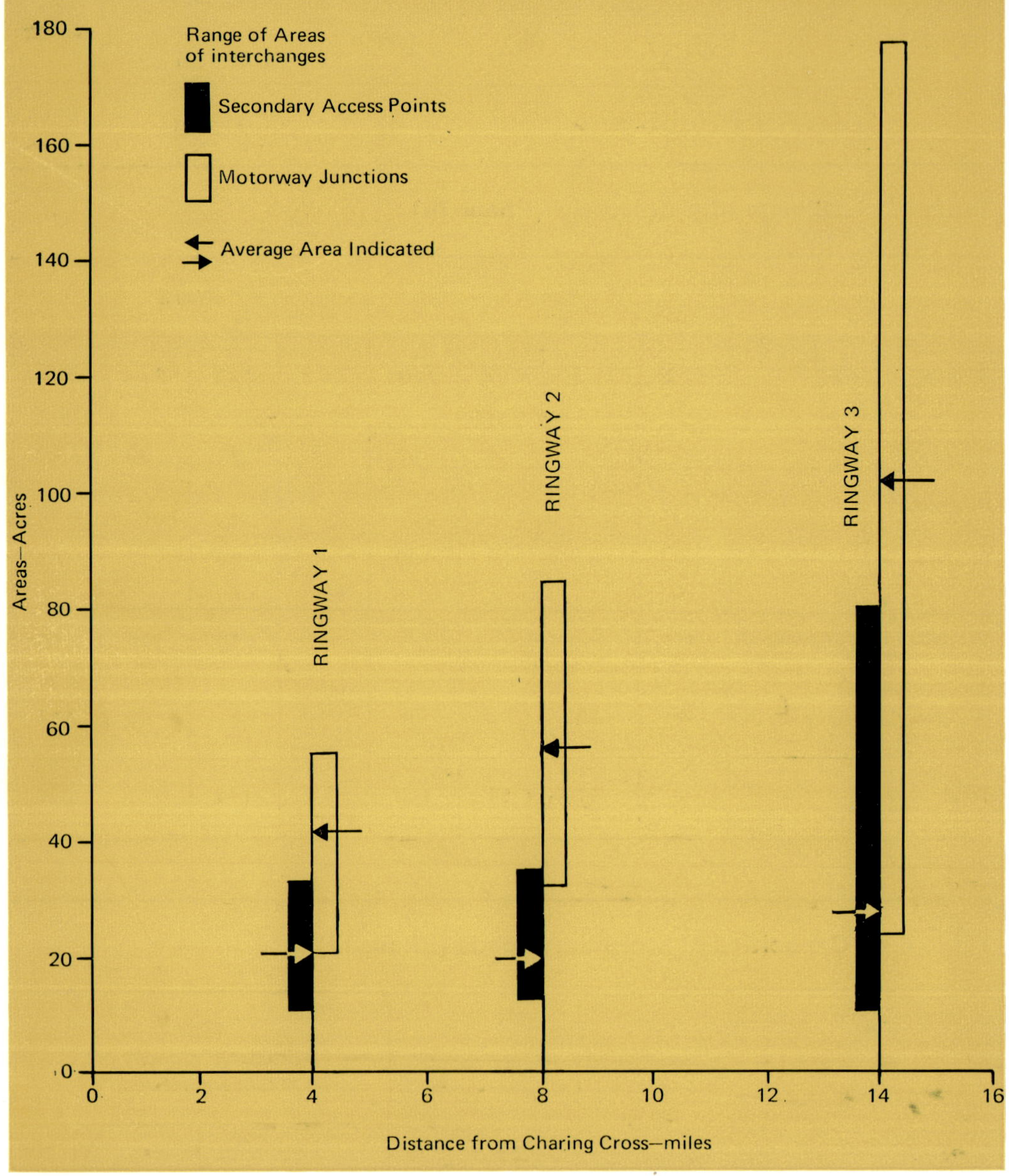

Table 4.5 Average spacing* of interchanges on motorways (Phase III)

Location	Plan 1: All interchanges	Plan 1: Local access points	Plan 3: All interchanges	Plan 3: Local access points
Ringway 1	1·5	1·5	1·3	1·7
Ringway 2	—	—	1·5	2·3
Ringway 3	2·8	3·6	2·9	5·5
Radials	1·9	1·9	1·5	1·9

*Spacing in miles

4.20 The frequency of interchanges adopted for plans 1 and 3 is shown in Table 4.5. Figure 4.5 compares the size of the interchanges on the system and illustrates the much greater land requirements of motorway-motorway interchanges as compared with motorway-secondary interchanges. Once again it should be noted that the interchange designs used were of a standard type according to the likely traffic volumes and in many cases may be higher in standard and cost than would actually be built when other factors are taken into account.

Public transport plans

4.21 Two public transport plans were tested after a number of alternatives had been considered. The first plan is illustrated in Figure 4.6 and was known as C2/B. It consisted of a British Rail plan C2 and an Underground plan B, together with a high speed bus system to form a composite public transport network. The constituent parts of the plan were:

a. *British Rail Plan C2*

This plan consisted of the 1962 rail network with the electrification of the Great Northern line to Royston and the Hertford Loop; the diversion of these services to Moorgate by the Northern City line (transferred from the London Transport Board); the electrification of the Lea Valley line, St. Pancras to Bedford line, Euston to Crewe line, Blackfriars–Farringdon and Kentish Town–Farringdon links to enable through running from north to south London; the opening of the West London line to a commuter service and a link from Victoria to London Airport. In addition Southern Region services would be speeded up by 15 per cent. This would be achieved by track improvements and increased motive power.

b. *Underground Plan B*

This plan consisted of the 1962 network plus the Victoria Line with the Brixton extension, the Fleet Line which would be the Stanmore Branch of the Bakerloo Line extended by a new line from Baker Street to Lewisham (via Bond Street, Green Park, Trafalgar Square, Aldwych, Ludgate, Fenchurch Street, Surrey Docks and New Cross) and the Piccadilly Line extended from Aldwych to Waterloo and from Hounslow West to London Airport.

c. *The Bus System*

This consisted of the 1962 bus and Green Line coach network, with the addition of motorway buses running both on motorways and through the Central Area, a network of standee (Red Arrow) buses in the Central Area and a network of buses linking points on Ringway 1. It was assumed for the purpose of the test that the new services could run at uncongested road speeds. The existing bus network was tested at normal

Figure 4.6 Rail improvements in plan C2/B (Phase III)

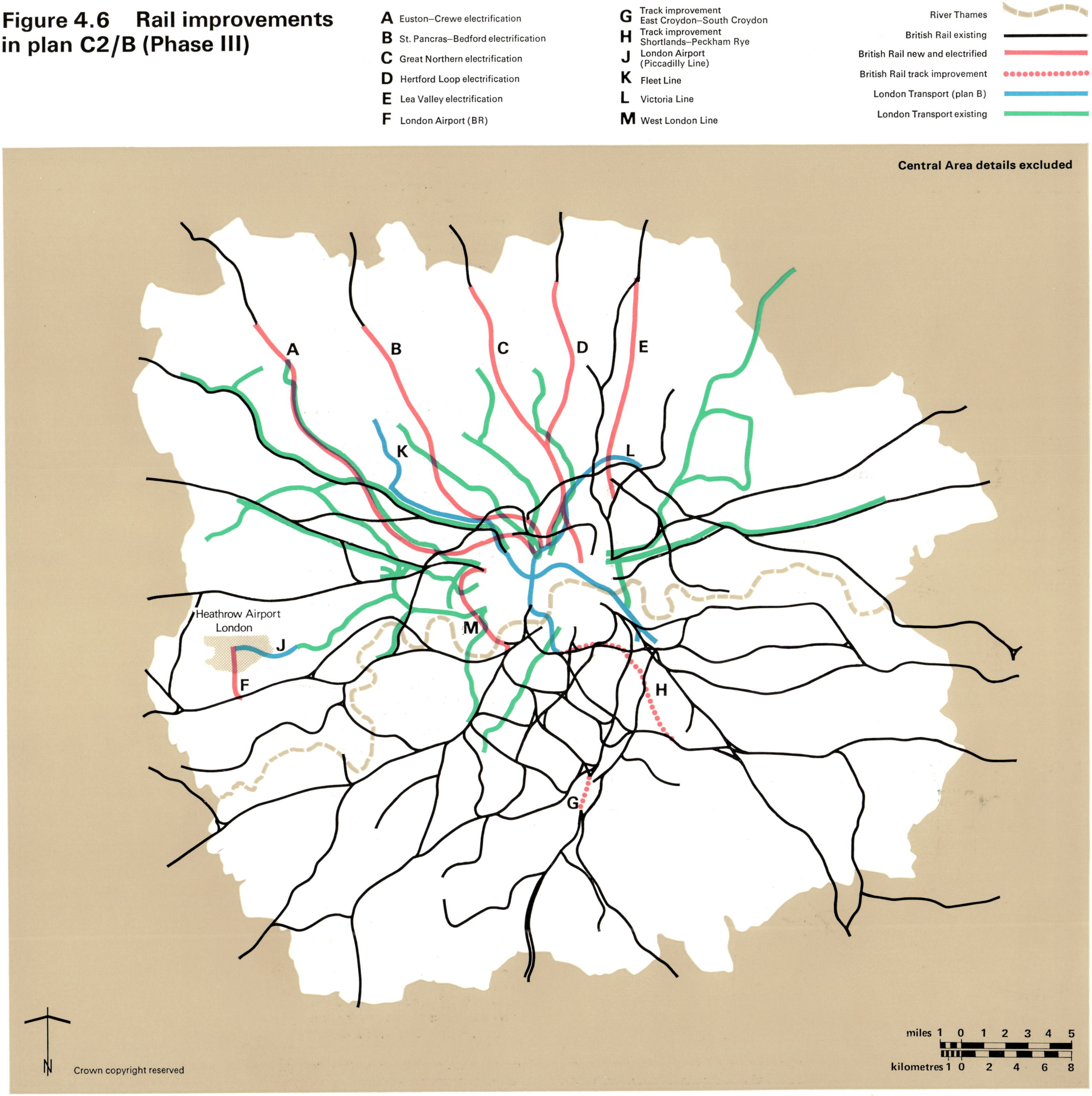

Figure 4.7 Rail improvements in Plan G/B —outer area (Phase III)

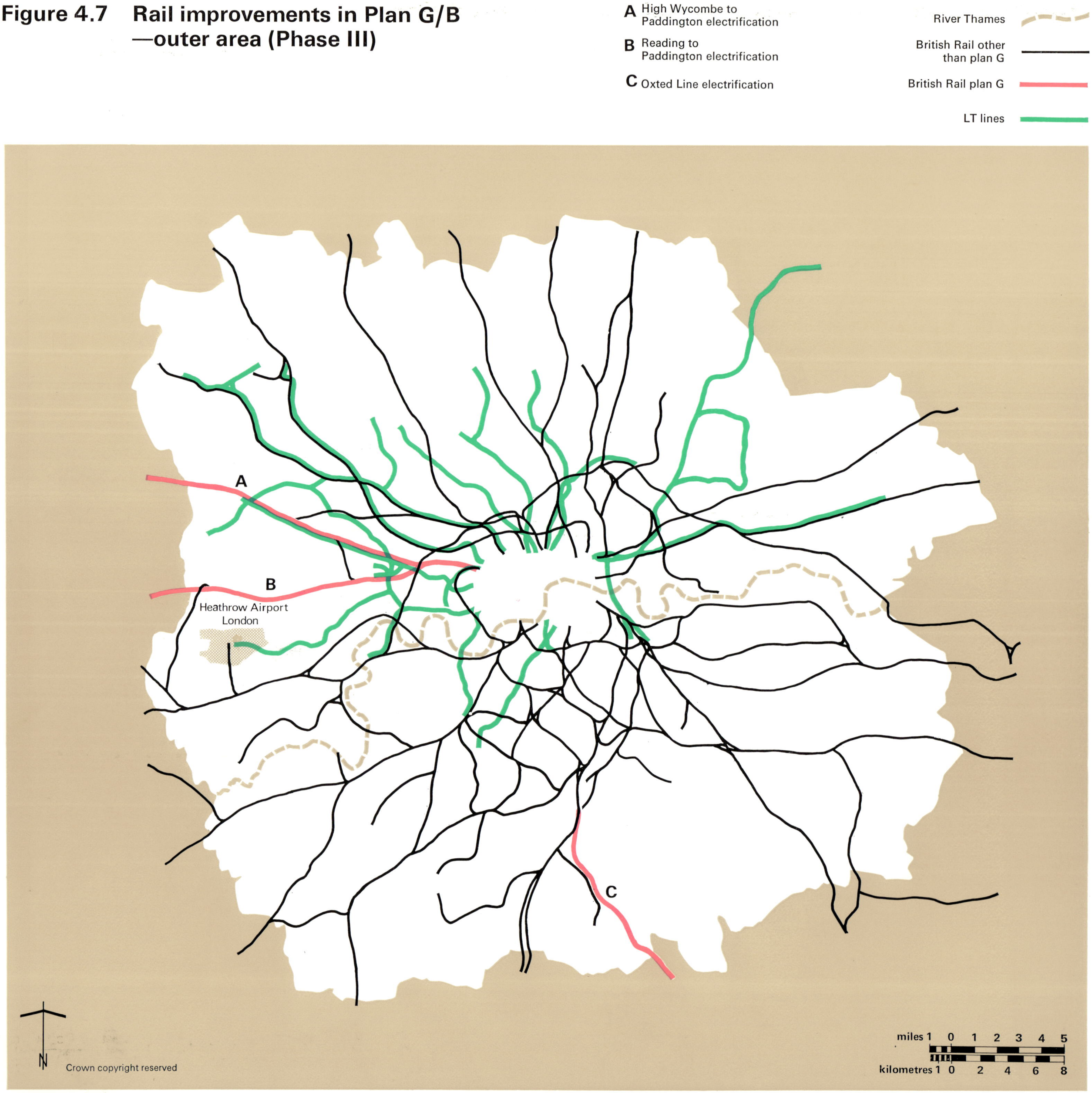

Figure 4.8 British Rail plan G/B Central Area cross links (Phase III)

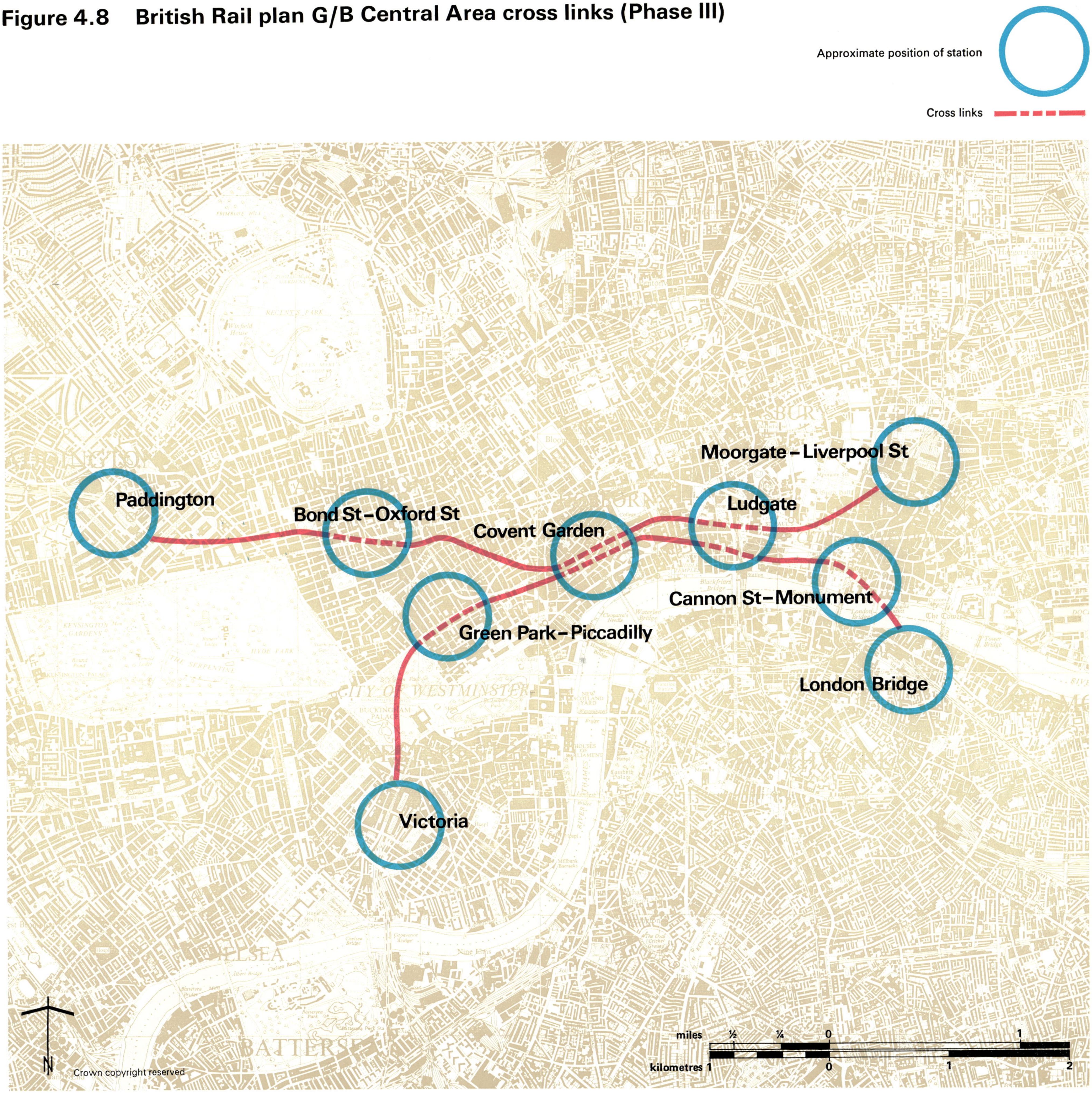

scheduled speed, excluding certain allowances for delay at particular points of congestion which had been assumed for 1962.

4.22 In seeking an alternative public transport plan for the comparative tests the Greater London Council and the Ministry of Transport were anxious that a quite different concept should be developed and Appendix B describes some of the plans prepared by the consultants. They concentrated on improving accessibility to the Central Area for British Rail passengers by providing main line services within the area bounded by the existing main line termini. British Rail and London Transport cooperated in preparing from the consultants' plans a plan that could be used for the purposes of the study. The resulting plan, designated G/B, merely represents hypothetical proposals for the purpose of testing in Phase III. Nevertheless it does contain ideas for improving public transport that are not infrequently suggested. Plan G/B comprises the elements of plan C2/B plus additional routes linking main line termini, Paddington to Liverpool Street and Victoria to London Bridge. The plan is shown in Figures 4.7 and 4.8.

4.23 It became apparent as the tests proceeded that the assumed provision of the high speed buses in the Central Area and on the motorways in plan C2/B, with speeds higher than those of the other traffic on the same roads, had led to trips being drawn off the Underground system in an unrealistic manner. Plan C2/B/1 was therefore prepared with these buses running at the final assignment speeds of plan 3 and with certain other small modifications. It is the assignments to plan C2/B/1 which have largely been used in the operational evaluation of the public transport tests, as it was found that the change in bus speeds did not significantly effect the total public transport matrix.

Environmental standards

4.24 In determining the location of the various parts of the primary road network for the purposes of preparing cost estimates it was necessary to establish geometric, structural and environmental standards for the use of the engineers and planners. Consistency throughout and between the road plans could then be obtained for the purposes of the economic analysis. The guidelines which were established are given in detail in Appendix D but the main points were as follows:

a. In residential areas the motorway should be concealed and normally depressed.

b. The edge of the pavement should be at least 60 feet from the nearest dwelling if possible.

c. Where the motorway was depressed the 60 feet clearance standards could be relaxed.

d. In areas of public buildings or urban open space the motorway should be concealed and normally depressed.

e. In shopping areas and industrial sites any form of construction could be used.

f. In rural areas the motorway would normally be at ground level.

These guidelines offer considerable protection to the environment in that they provide for a large clearance between the motorway and the nearest dwellings. This did lead to cost estimates which were higher than usual practice though no estimates of the sensitivity of the costs to variations in the standards was obtained.

Network costing

4.25 As part of the study it was necessary to examine the alignment and routing of both the public transport and the primary road networks to determine the capital costs. Table 4.6 gives, for lengths of motorway between interchanges only, the area required per 1,000 feet for different widths of road together with the costs per foot run. Table 4.7 summarises the differences between the complete plans 1, 3 and 9. The cost of land for plan 3 is nearly nine times greater than for plan 1, the cost of construction less than four times and the area of land consumed approximately double. These figures reflect the fact that the majority of the routes in plan 1 ran through rural areas on the outskirts of London while in plan 3 there were many more routes and they were much more heavily concentrated within the built-up area. More details of the costing procedure and the results for plan 3 are given in Appendix C.

4.26 In addition to the direct costs of the motorways, estimates were made of costs of extra parking spaces that would be required in the Central Area and the former LCC area to cater for the increased amount of movement being planned for and these figures are also given in Table 4.7. Maintenance costs were estimated on an annual basis so as to be set against annual benefits in the economic evaluation.

4.27 For the public transport plans only the incremental costs of plan G/B over C2/B were calculated. The main difference in cost between these plans was the bored tunnels through central London and the electrification of the Reading, High Wycombe and Oxted lines. Owing to the tentative nature of the central London links from British Rail's viewpoint the cost estimates are of a very general nature. The capital costs assumed a tunnel bore of 19ft. 4 in. To provide an order of magnitude of the incremental capital costs involved they are shown in Table 4.8. The remaining costs calculated on the public transport side consisted of increased rail operating costs, increased engineering, depreciation and renewal costs and a decrease in the operating costs of buses where traffic had been diverted from them to the Central Area links. Again it should be noted that the operating costs of the deep level tunnels were based upon broad assumptions. The general magnitude of these costs is shown in Table 4.9.

Phase III results

Introduction

4.28 The model developed for Phase III and the method of economic evaluation represent considerable advances

Table 4.6 Unit construction costs* for plan 3 (Phase III)

		Dual-4		*Dual-3*		*Dual-2*	
Location	*Construction*	*Area acres/1000'*	*Cost £/ft*	*Area acres/1000'*	*Cost £/ft*	*Area acres/1000'*	*Cost £/ft*
Ringway 1	E(V)	5·46	1,245	—	—	—	—
	E(R)	—	—	—	—	—	—
	E	—	—	—	—	—	—
	G	4·23	185	—	—	—	—
	D	—	—	—	—	—	—
	D(R)	4·47	1,443	—	—	—	—
	T(C)	4·99	2,747	—	—	—	—
Inner radial	E(V)	4·97	1,310	4·65	885	—	—
	E(R)	—	—	—	—	—	—
	E	—	—	—	—	—	—
	G	4·06	163	5·44	197	—	—
	D	—	—	5·45	127	—	—
	D(R)	3·79	1,403	4·51	986	—	—
	T(C)	3·28	2,027	5·32	1,698	—	—
Ringway 2	E(V)	4·52	1,036	2·80	841	—	—
	E(R)	5·08	291	—	—	—	—
	E	3·80	173	—	—	—	—
	G	4·58	174	5·09	104	—	—
	D	5·93	130	4·15	158	—	—
	D(R)	4·94	843	5·54	1,076	—	—
	T(C)	4·45	2,076	3·06	1,575	—	—
Outer radial	E(V)	4·64	1,054	4·54	888	—	—
	E(R)	6·03	300	4·72	301	—	—
	E	4·52	235	—	—	—	—
	G	5·30	141	5·09	124	4·58	63
	D	5·22	195	—	—	—	—
	D(R)	5·69	928	5·06	889	—	—
	T(C)	—	—	—	—	—	—
Ringway 3	E(V)	3·96	1,078	3·70	759	—	—
	E(R)	—	—	5·20	215	—	—
	E	—	—	5·00	117	—	—
	G	5·69	169	5·23	106	—	—
	D	—	—	5·11	92	—	—
	D(R)	—	—	5·11	796	—	—
	T(C)	—	—	—	—	—	—

Definitions: E(V)—Elevated (Viaduct) E(R)—Elevated (Retained) E—Embankment G—Ground level D—Depressed (open cut) D(R)—Depressed (Retained) T(C)—Tunnel (Cut and cover)

*For lengths of motorway between interchanges only

Table 4.7 Road network costs* (Phase III)

	Plan 1	*Plan 3*	*Plan 9*
Route length (miles)	150	347	413
Area of land (acres)	4,651	10,753	13,082
Cost of land	59	513	637
Cost of construction	313	1,145	1,426
Motorway cost	372	1,658	2,063
Cost of parking	79	183	213
Total cost	451	1,841	2,276

*Costs in £ million (at 1966 prices)

Table 4.8 Incremental capital costs of plan G/B over C2/B (Phase III)

Item	£(*m*)
Bored tunnels	141·0
End ramps and connections	6·5
Land and compensations	1·5
Electrification to Reading, High Wycombe and Oxted	15·0
Contingencies	11·0
Total	175·0

over what had been achieved in Phase II of the LTS. Nevertheless before examining the results obtained it is necessary to bear in mind certain facts about the model and input data.

4.29 The Phase III forecasts were made for the year 1981. Two of the three road plans tested in order to examine the implications of a wide range of investment involved greater capital sums than are likely to be expended by that date. The two public transport networks on the other hand are more capable of achievement if it were decided that they are desirable.

4.30 In common with most other transportation studies the LTS calculated modal choice on the basis of travel time by private and public transport. To this extent the behavioural assumptions were not strictly correct since they ignored travel cost, comfort and convenience, although of course these factors are implicit in the 1962 data which were used to calibrate the modal choice part of the model.

4.31 For both public transport and private trips the assignment procedure was on the basis of minimum time path calculations. For the road networks it may be that certain of the loadings are in error particularly where all the trips from one traffic district enter the road network. Thus it is the general pattern of over- and under-loading which should receive attention rather than the situation on individual links in the road networks. The consultants considered the likely effect on the economic evaluation of using minimum cost rather than minimum time assignments and concluded that they would not produce substantially different results. Similar comments apply to the public transport networks but in addition it should be noted that no account was taken of the nuisance value of interchange as distinct from the time taken in interchange, which was allowed for. To the extent that this affects public transport users separately from time the techniques will have underestimated the value of the cross London links in plan G/B.

4.32 All of the forecasts of road trips were carried out on a 24-hour basis. It

Table 4.9 Differences in annual operating costs between plans G/B and C2/B (Phase III)

Item	£(*m*)
Operational—rail	1·12
Engineering—rail	·73
General—rail	·19
Depreciation and renewals—rail	·13
Increase in rail operating costs	2·17
Decrease in bus operating costs	2·54
Net saving in operating costs	0·37

Figure 4.9 Plans tested in Phase III

Plans	C2/B	G/B
1	Test 2 (BPR)	—
3	Test 1 (BPR) Test 4 (Transitnet)	Test 5 (Transitnet)
9	Test 3 (BPR)	—

is possible to work out peak hour loadings by factoring these results. However it is clearly desirable to extend the work to make specific peak hour forecasts and this will be done in future.

4.33 Finally, mention has already been made of the fact that the population, employment and income forecasts in the LTS are now considered to be high in the light of work carried out very recently for the Greater London Development Plan. The consultants investigated the sensitivity of road traffic volumes to changes in population and employment as described later in paragraph 4.60. The Council has already completed some work with revised population, employment and income forecasts as reported in Chapter 5 and is continuing work with further revisions.

4.34 During Phase III five tests were carried out, three road tests with the same public transport network and two public transport tests with the same road network. In the road tests the public transport network was plan C2/B whilst in the public transport tests the road network was plan 3. This report presents the results for all five tests. Figure 4.9 shows the five tests and indicates that two tests were carried out with plans 3 and C2/B, a road test using the Bureau of Public Roads public transport assignment programme and a public transport test using Transitnet (see paragraph 4.8).

Results

4.35 The number of trips by mode for each of the five tests is shown in Table 4.10. They are divided by the three purposes of work, other home-based and non-home-based. Commercial and taxi trips and external trips are also shown. The difference between tests 1 and 4 results from the different public transport assignment procedures, Bureau of Public Roads and Transitnet respectively, which had an effect on the distribution, modal choice and assignment stages.

4.36 Prior to restraint there were virtually the same number of internal driver trips for the three road tests (1, 2 and 3) and the same number of public transport trips from the modal choice calculation. The restraint procedure had a very different effect in each case however. In the road test of plan 3 there were 0·8 million trips removed from the road network by restraint. Of these, 0·32 million transferred to public transport, leaving 0·48 million which were not made at all. The corresponding figures for plan 1, which represented the low level of road investment, were very much greater—2·89 million, 1·14 million and 1·75 million respectively. No restraint was necessary for plan 9.

4.37 The number of public transport trips from non-car-owning households was the same for each of the five tests, 1·77 million. The number of trips in 1981 which were equivalent to the unreported trips in 1962 were assumed to be the same for all five tests, namely 0·82 million. The commercial, taxi and external trip totals were not varied with changes in the network except in the case of plan 1, where they were reduced in proportion to trip length ratios between plans 1 and 3. The number of external public transport trips was constant throughout at 0·85 million.

4.38 The difference in trip totals between the two public transport tests (4 and 5) was negligible, which suggests that the trips which might be expected to use the cross-London links in plan G/B would transfer within public transport modes and not from private transport. The difference which results from the use of the Transitnet and Bureau of Public Roads public transport assignment procedures is shown in the results of tests 4 and 1, both for plans 3 and C2/B. Before restraint there is a small difference in private and commercial trips but this has virtually disappeared after restraint. The difference remains in the public transport trip totals however with Transitnet (test 4) resulting in a higher estimate of public transport trips than BPR (test 1) by some 150,000 trips.

Table 4.10 Summary of trips* (Phase III)

	Private and commercial driver		*Public transport passenger*				
	Pre-restraint	*Post-restraint*	*NCO households*	*Mode choice*	*Restraint*	*Unreported*	*Total*
Test (plans)	***Test 1 (3 and C2/B)***						
Work	2·42	2·17	1·16	1·73	0·19	0·21	3·29
Other home-based	4·16	3·76	0·54	1·71	0·11	0·19	2·55
Non-home-based	1·57	1·42	0·07	0·18	0·02	0·42	0·69
Internal driver	8·15	7·35					
Commercial and taxis	1·94	1·94					
Total internal	10·09	9·29	1·77	3·62	0·32	0·82	6·53
External	1·22	1·22					0·85
Total	11·31	10·51					7·38
Test (plans)	***Test 2 (1 and C2/B)***						
Work	2·39	1·50	1·16	1·77	0·66	0·21	3·80
Other home-based	4·14	2·71	0·54	1·73	0·39	0·19	2·84
Non-home-based	1·56	0·99	0·07	0·19	0·09	0·42	0·77
Internal driver	8·09	5·20					
Commercial and taxis	1·63	1·63					
Total internal	9·72	6·83	1·77	3·69	1·14	0·82	7·41
External	1·03	1·03					0·85
Total	10·75	7·86					8·26
Test (plans)	***Test 3 (9 and C2/B)***						
Work	2·43	NO RESTRAINT	1·16	1·73	—	0·21	3·10
Other home-based	4·16		0·54	1·71	—	0·19	2·43
Non-home-based	1·57		0·07	0·18	—	0·42	0·67
Internal driver	8·16						
Commercial and taxis	1·94						
Total Internal	10·10		1·77	3·62	—	0·82	6·20
External	1·22						0·85
Total	11·32						7·05
Test (plans)	***Test 4 (3 and C2/B)***						
Work	2·29	2·11	1·16	1·87	0·13	0·21	3·37
Other home-based	4·11	3·80	0·54	1·76	0·08	0·19	2·57
Non-home-based	1·52	1·41	0·07	0·23	0·02	0·42	0·74
Internal driver	7·92	7·32					
Commercial and taxis	1·94	1·94					
Total internal	9·86	9·26	1·77	3·86	0·23	0·82	6·68
External	1·22	1·22					0·85
Total	11·08	10·48					7·53
Test (plans)	***Test 5 (3 and G/B)***						
Work	2·29	2·11	1·16	1·87	0·13	0·21	3·37
Other home-based	4·11	3·82	0·54	1·76	0·08	0·19	2·57
Non-home-based	1·52	1·41	0·07	0·23	0·02	0·42	0·74
Internal driver	7·92	7·34					
Commercial and taxis	1·94	1·94					
Total internal	9·86	9·28	1·77	3·86	0·23	0·82	6·68
External	1·22	1·22					0·85
Total	11·08	10·49					7·53

*24-hour average weekday millions of trips

Where the results of Phase III are compared with those for 1962 and for Phase II later in this chapter, it is the results of test 4 after restraint which are used.

4.39 Tables 4.11 and 4.12 show the summaries of vehicle-miles and vehicle-hours for the private and commercial trips assigned to the networks. The trip totals on which Tables 4.11 and 4.12 are based are less than those in Table 4.10 because they exclude all intra-district movements which are not assigned to the network. Thus for test 1 there were a total of 10·51 —7·87 =2.64 million intra-district trips not assigned. The results in Tables 4.11 and 4.12 support those in Table 4.10 in that they show that plan 3 carries slightly less travel after restraint than plan 9 which is not restrained. Both the vehicle-miles and vehicle-hours for plan 1 are substantially less than for any of the other plans while the results for the two public transport alternatives are substantially the same as each other.

4.40 Tables 4.13 and 4.14 show the passenger-miles and passenger-hours for the public transport trips assigned to the networks. For tests 4 and 5 the results are shown for both 24-hours and peak hour. Peak hour results are also shown for the additional test of plan C2/B/1 which had high speed buses running at the restrained road speeds rather than running freely.

4.41 Since the public transport assignments were carried out between zones, that is areas smaller than districts, the full totals of trips shown in Table 4.10 are included. The results are divided between British Rail and London Transport (Underground and bus) trips. The results confirm again the small difference in total travel between the public transport plans although it should be noted that no attempt was made to examine the effect of local congestion on trip patterns and totals as was done for road trips.

4.42 The passenger-mile totals are of the same order for British Rail and London Transport trips but, because of the higher average speed for British Rail trips, passenger-hour totals are much

Table 4.11 Vehicle-miles summary for private and commercial trips* (Phase III)

Plans	Restraint	Trips	*Vehicle-miles* Primary	Secondary	Total
Test 1 3 and C2/B	Pre-	8·43	34·38	28·86	63·24
	Post-	7·87	32·10	26·95	59·05
Test 2 1 and C2/B	Pre-	7·91	13·31	40·35	53·66
	Post-	5·88	11·49	30·51	42·00
Test 3 9 and C2/B	—	8·44	38·17	25·95	64·12
Test 4 3 and C2/B	Pre-	8·19	33·37	27·58	60·95
	Post-	7·81	31·81	26·30	58·11
Test 5 3 and G/B	Pre-	8·19	33·32	27·55	60·87
	Post-	7·80	31·74	26·24	57·98

*24-hour average weekday millions of vehicle trips and miles

Table 4.12 Vehicle-hours summary for private and commercial trips* (Phase III)

Plans	Restraint	Trips	*Vehicle-hours* Primary	Secondary	Total
Test 1 3 and C2/B	Pre-	8·43	0·85	1·30	2·15
	Post-	7·87	0·80	1·21	2·01
Test 2 1 and C2/B	Pre-	7.91	0·28	1·75	2·03
	Post-	5·88	0·24	1·30	1·54
Test 3 9 and C2/B	—	8·44	0·93	1·17	2·10
Test 4 3 and C2/B	Pre-	8·19	0·83	1·23	2·06
	Post-	7·81	0·79	1·18	1·97
Test 5 3 and G/B	Pre-	8·19	0·83	1·23	2·06
	Post-	7·80	0·79	1·17	1·96

*24-hour average weekday millions of vehicle trips and hours

Table 4.13 Passenger-miles summary for public transport trips* (Phase III)

Plans		*Trips*	*Passenger-miles* *BR*	*LTB*	*Total*
Test 4					
3 and C2/B	Peak	1·11	4·74	4·21	8·95
	24-hour	7·53	27·79	28·39	56·18
3 and C2/B1	Peak	1·11	4·98	4·02	9·00
Test 5					
3 and G/B	Peak	1·11	4·97	3·96	8·93
	24-hour	7·53	28·92	27·21	56·13

*24-hour average weekday millions of passenger trips and miles

Table 4.14 Passenger-hours summary for public transport trips* (Phase III)

Plans		*Trips*	*Passenger-hours* *BR*	*LTB*	*Motorised total*	*Walking, waiting*	*Total*
Test 4							
3 and C2/B	Peak	1·11	0·13	0·24	0·37	0·30	0·67
	24-hour	7·53	0·75	1·63	2·38	1·99	4·37
3 and C2/B/1	Peak	1·11	0·14	0·24	0·38	0·31	0·69
Test 5							
3 and G/B	Peak	1·11	0·14	0·23	0·37	0·30	0·67
	24-hour	7·53	0·80	1·57	2·37	1·98	4·35

*24-hour average weekday millions of passenger trips and hours

less than for London Transport. The results also illustrate the substantial amount of time spent walking and waiting during public transport trips. Similar information is not available for private car trips.

4.43 Tables 4.15 and 4.16 show for plans 1 and 3 the pattern of private and commercial vehicle movements in more detail. Sector to sector movements are shown for all internal vehicle trips before and after restraint. Table 4.17 shows internal and external 24-hour and peak hour public transport passenger movements between sectors on plan C2/B after the restraint procedure for road travel.

Comparison of results from Phases I, II and III

4.44 The previous section has compared the main results of the five tests in Phase III. It is helpful to compare these results with those for 1962 and 1981 given by Phase I and II of the LTS. In order to do this only one test is used to represent Phase III, that is test 4 after restraint. It is this test which has a road network most similar to the 1981A network of Phase II.

4.45 Table 4.18 shows the total internal person movements for each of the three phases. There is an increase of more than 5 million trips between 1962 and 1981, for both Phases II and III. The split between modes is quite different however because of the procedures used in Phase III. Very broadly there are 1·5 million internal private trips fewer in the Phase III results and 1·5 million more public transport trips. There is still a large increase in the total number of private trips, some 90 per cent more than 1962, but in Phase III the total number of public transport trips does not fall below the 1962 total.

4.46 Table 4.19 shows for the three phases a summary of vehicle-miles and vehicle-hours on the road networks and of passenger-miles and passenger-hours on the public transport networks.

Table 4.15 Total internal vehicle sector to sector trips* (Phase III)—Test 2 (Plans 1 and C2/B)

Generation sectors	*Attraction sectors* 0	1	2	3	4	5	6	7	8	9	*Totals*
0	1,305,510	84,513	50,387	10,661	90,926	69,854	3,212	20,116	5,185	108,525	1,748,889
	806,954	53,212	31,620	7,628	63,028	38,913	1,961	15,334	4,778	95,561	1,118,989
1	72,110	829,624	30,938	39,693	89,604	69,075	3,257	11,885	3,708	8,522	1,158,416
	46,268	515,436	19,349	28,329	61,019	37,561	1,886	8,790	3,390	7,714	729,742
2	41,837	31,440	233,695	1,902	4,930	6,379	302	1,543	297	1,931	324,256
	26,144	19,222	141,983	1,461	3,637	3,708	217	1,273	292	1,801	199,738
3	15,299	55,642	3,596	923,372	79,102	41,373	3,171	24,369	24,035	7,868	1,177,827
	10,056	37,072	2,356	580,846	54,474	23,605	1,966	17,728	21,119	7,163	756,385
4	94,570	89,394	8,828	62,091	610,324	226,332	10,883	54,154	8,509	27,701	1,192,786
	63,146	59,432	5,790	43,308	423,832	137,285	6,791	41,038	7,920	25,213	813,755
5	30,325	22,265	3,803	18,765	123,662	331,707	21,490	46,216	7,081	15,082	620,396
	20,853	15,313	2,540	13,228	96,693	256,701	17,307	35,343	6,403	13,737	478,118
6	1,805	1,224	212	1,920	6,477	22,522	9,195	13,432	1,121	1,802	59,710
	1,325	895	162	1,423	5,106	17,859	7,470	10,806	1,041	1,692	47,779
7	29,659	11,814	2,967	33,769	68,509	115,783	19,846	680,194	81,662	105,510	1,149,713
	20,724	8,101	2,017	22,352	48,673	65,805	12,663	457,476	72,878	93,744	804,433
8	7,178	4,281	691	21,589	10,749	33,210	2,680	87,136	532,713	37,920	738,147
	4,777	2,772	465	13,440	7,727	17,741	1,527	58,427	466,707	33,548	607,131
9	99,347	8,276	5,234	9,226	37,999	66,889	4,364	104,931	31,896	1,187,650	1,555,812
	62,526	5,439	3,274	6,163	26,072	36,013	2,533	69,952	28,349	1,039,534	1,279,855
Totals	1,697,640	1,138,473	340,351	1,122,988	1,122,282	983,124	78,400	1,043,976	696,207	1,502,511	9,725,952
	1,062,773	716,894	209,556	718,178	790,261	635,191	54,321	716,167	612,877	1,319,707	6,835,925

*24-hour average weekday internal vehicle trips: upper figure—pre-restraint; lower figure—post restraint

Table 4.16 Total internal vehicle sector to sector trips* (Phase III)—Test 4 (Plans 3 and C2/B)

Generation sectors	*Attraction sectors*										
	0	1	2	3	4	5	6	7	8	9	*Totals*
	1,284,726	82,282	46,143	16,658	101,946	64,549	3,243	26,875	7,765	112,458	1,746,645
	1,071,431	75,785	35,714	16,316	95,840	64,539	3,237	26,444	7,762	110,750	1,507,818
1	69,365	817,844	30,358	46,652	97,653	71,305	3,639	13,479	5,083	10,614	1,165,992
	57,627	712,186	23,132	44,169	95,566	71,299	3,631	13,391	5,081	10,379	1,036,461
2	39,492	33,080	229,960	2,361	6,327	8,387	393	1,993	643	2,077	324,713
	33,227	29,809	173,958	2,310	6,084	8,387	392	1,984	642	2,053	258,846
3	20,346	61,358	4,043	916,923	79,420	40,593	2,600	27,133	29,557	9,951	1,191,924
	16,837	52,904	3,165	893,077	78,538	40,587	2,596	27,089	29,529	9,844	1,154,166
4	110,317	104,207	12,454	66,279	612,983	201,837	8,824	57,680	10,098	30,367	1,215,046
	92,387	91,398	9,536	65,205	597,156	201,804	8,813	57,061	10,093	29,609	1,163,062
5	33,029	26,933	4,868	21,037	129,934	353,923	22,734	43,606	7,005	15,173	658,242
	28,194	24,266	3,800	20,705	127,566	353,868	22,699	43,448	6,999	14,984	646,529
6	2,112	1,537	269	1,814	6,460	24,034	10,218	14,354	1,164	2,002	63,964
	1,873	1,399	228	1,795	6,408	24,032	10,195	14,330	1,164	1,987	63,411
7	44,592	14,944	4,830	38,640	70,565	81,159	18,430	701,630	83,438	111,679	1,169,907
	37,856	13,406	3,709	38,363	69,193	81,141	18,399	696,939	83,390	109,525	1,151,921
8	10,204	6,064	1,606	27,345	12,597	29,507	2,560	90,037	530,978	36,736	747,634
	8,536	5,281	1,215	27,103	12,298	29,506	2,556	89,739	530,609	36,509	743,352
9	102,606	10,393	5,558	12,041	40,802	57,654	4,070	111,554	31,513	1,196,751	1,572,942
	85,810	9,396	4,261	11,945	39,030	57,649	4,059	109,060	31,486	1,187,498	1,540,194
Totals	1,716,789	1,158,642	340,089	1,149,750	1,158,687	932,948	76,711	1,088,341	707,244	1,527,808	9,857,009
	1,433,778	1,015,830	258,718	1,120,988	1,127,679	932,812	76,577	1,079,485	706,755	1,513,138	9,265,760

*24-hour average weekday internal vehicle trips: upper figure—pre-restraint; lower figure—post-restraint

Table 4.17 Total public transport passenger sector to sector trips* (Phase III)—Test 4 (Plans 3 and C2/B)

Generation sectors	Attraction sectors: 0	1	2	3	4	5	6	7	8	9	*External*	*Totals*
0	518,037	26,140	16,168	1,765	53,406	193,016	8,897	12,995	136	32,235	5,334	868,129
	72,287	2,760	3,347	31	8,094	46,710	2,280	869	2	3,903	166	140,449
1	34,024	341,798	10,038	12,021	53,780	180,506	9,036	6,680	118	4,944	3,675	656,620
	3,467	43,192	1,608	828	8,267	43,754	2,449	145	1	203	109	104,023
2	16,738	7,492	101,255	241	2,814	21,443	728	311	5	214	1,447	152,688
	2,652	748	13,002	3	296	6,603	199	90	0	0	45	23,638
3	10,596	31,805	3,369	425,462	60,946	227,005	12,471	18,013	10,520	4,199	6,800	811,186
	641	1,592	203	46,316	6,554	54,025	3,526	1,147	652	54	211	114,921
4	108,327	72,065	11,830	40,971	381,141	542,730	38,095	39,811	14,877	28,171	4,621	1,282,639
	7,649	2,762	1,432	2,208	49,037	89,606	4,715	2,070	1,130	2,381	143	163,133
5	37,956	16,083	5,434	16,173	88,161	314,224	19,553	29,357	14,073	30,543	2,276	573,833
	2,999	901	337	620	5,625	20,687	1,222	2,066	637	1,528	71	36,693
6	2,653	1,030	433	1,070	4,923	23,244	3,228	5,680	2,404	5,520	1,100	51,285
	141	26	15	31	227	2,023	120	585	124	250	34	3,576
7	38,901	17,811	2,508	36,864	70,794	409,568	49,136	435,762	36,157	85,588	9,358	1,192,447
	4,528	1,489	140	2,183	11,260	83,246	9,611	56,466	3,165	8,325	285	180,698
8	3,694	757	407	22,132	8,046	133,994	9,596	29,152	157,703	11,433	1,531	378,445
	34	4	0	2,698	363	29,174	1,317	2,009	26,930	508	49	63,086
9	34,358	2,479	1,267	1,488	22,679	218,951	15,709	42,308	7,040	400,843	5,822	752,944
	4,876	92	22	26	1,808	59,916	5,128	5,086	698	65,124	194	142,970
External	10,879	8,548	4,215	24,099	56,383	627,141	39,644	17,129	3,833	15,485	0	807,356
	1,490	1,131	719	3,283	8,276	107,378	7,160	2,185	345	2,096	0	134,063
Totals	816,163	526,008	156,924	582,286	803,073	2,891,822	206,093	637,198	246,866	619,175	41,964	7,527,572
	100,764	54,697	20,825	58,227	99,807	543,122	37,727	72,718	33,684	84,372	1,307	1,107,250

*24-hour average weekday person trips: upper figure—24-hour total; lower figure—peak hour total

Table 4.18 Total internal person trips* (Phases I, II and III)

	Phase I 1962	*Phase II* 1981	*Phase III* 1981
Internal car driver	4·12	8·98	7·33
Other internal private	1·49	3·30	3·30
Total internal private	5·61	12·28	10·63
Internal public transport	5·72	4·41	5·86
Unreported public transport	0·72	0·82	0·82
Total internal public transport	6·44	5·23	6·68
Total internal	12·05	17·51	17·31

*24-hour average weekday millions internal trips

Table 4.19 Summary of total travel mileage and time* (Phases I, II and III)

	Phase I 1962	*Phase II* 1981	*Phase III* 1981
Private and commercial 24-hour trips	4·23	9·54	7·81
Vehicle-miles	24·95	60·25	58·11
Vehicle-hours	1·07	2·04	1·97
Public transport 24-hour trips	—	—	7·53
Passenger-miles	—	—	56·18
Passenger-hours	—	—	2·37
Public transport peak hour trips	1·10	—	1·11
Passenger-miles	7·63	—	9·00
Passenger-hours	0·36	—	0·38

*24-hour average weekday millions of vehicle and passenger trips, miles and hours
Private and commercial figures—district assignment; public transport figures—zonal assignment

Table 4.20 Comparison of modal choice of potential drivers*

	Phase I 1962	*Phase II* 1981	*Phase III* 1981
Driver	4·12	8·97	7·32
Public transport passenger	1·96	2·36	4·08
Total potential driver	6·08	11·33	11·40
Percent driver	68%	79%	64%

*24-hour average weekday millions internal trips (excludes private passenger)

In Table 4.19 the figures for the road networks for Phases I and II are based on assignments to district networks in order to be consistent with the results of Phase III. They do not therefore correspond with previously published results which were based on assignment to a zonal network.

4.47 The total number of potential car driver trips, ignoring car passengers, is made up in Table 4.10 of internal driver private trips plus those public transport trips from car-owning households which are calculated by the mode choice and restraint procedures. The proportion of these trips actually made as car driver trips in Phase III is shown in Table 4.20 and compared with equivalent proportions for Phase I and II. The Phase II results showed an increase in this proportion; in other words an increase in the use of the car by car-owners. The Phase III results on the other hand show a fall from 68 per cent in 1962 to 64 per cent in 1981. This is a significant result for it implies that in spite of the substantial investment in roads for plan 3, because of the very large increase in car ownership, the individual car owner would have less freedom to use his car in 1981 than in 1962. The corresponding proportion for plan 1 is 52 per cent which would represent a substantial decrease in the freedom of a car owner to use his car.

4.48 In Table 4.21 are shown the proportion of the internal car driver and public transport passenger trips by trip purpose for 1962 and for Phase III in 1981. Also shown are similar proportions for Phase III when car passengers are included. The results indicate a substantial proportional shift from the work to other home-based trip purpose for car driver trips and a similar but smaller shift for public transport trips.

Operational evaluation

4.49 There is no single figure which represents the capacity of a road although it is necessary to assume some figure in order to undertake traffic analyses. In fact different figures for capacity are associated with different levels of service on the road corresponding to free flow, congestion and complete traffic breakdown. In the LTS the consultants defined:

Figure 4.10 Overloaded and underloaded links on plan 3 (Phase III)

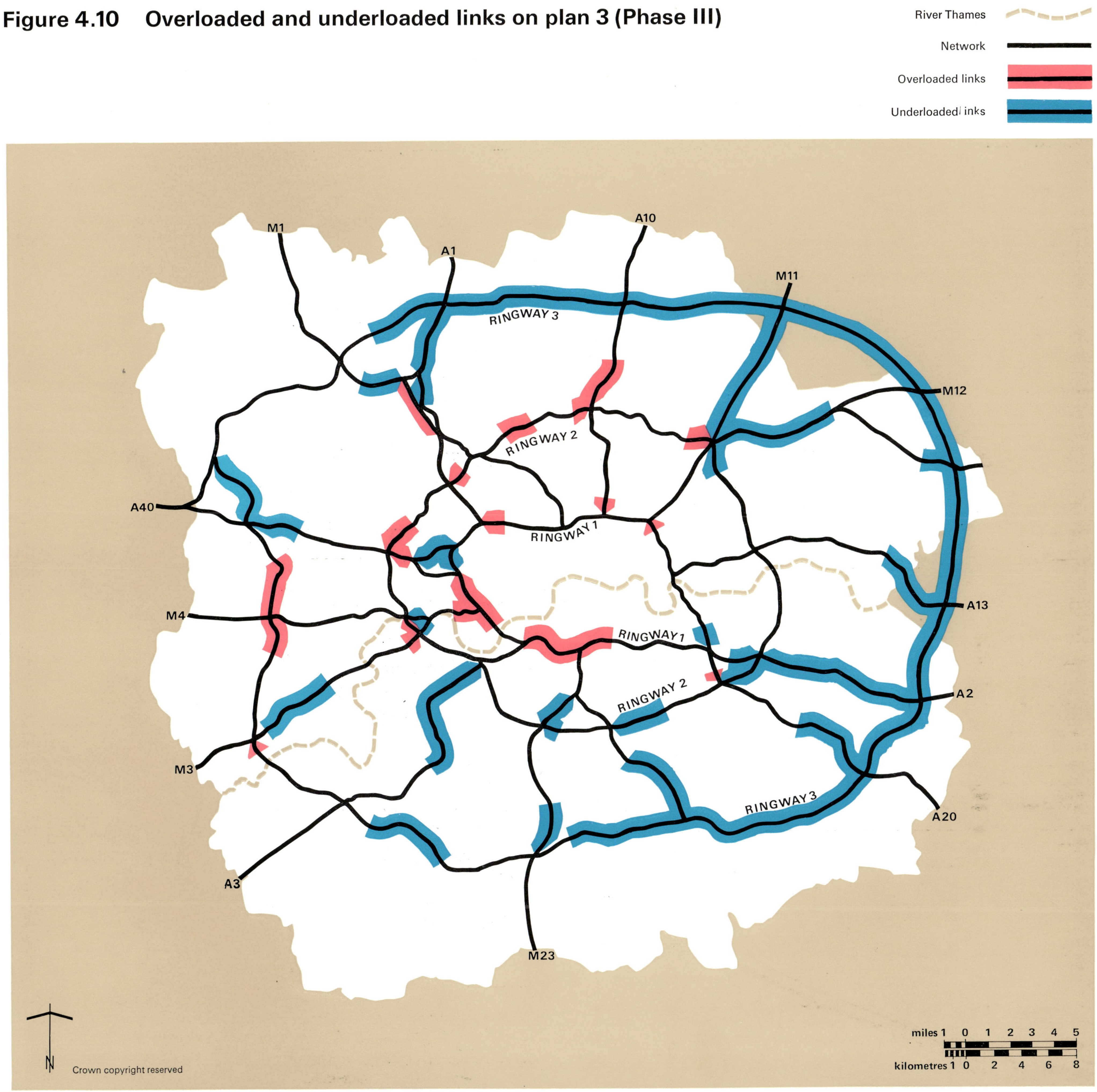

Figure 4.11 Overloaded Underground links—1981 matrix on 1962 network (Phase III)

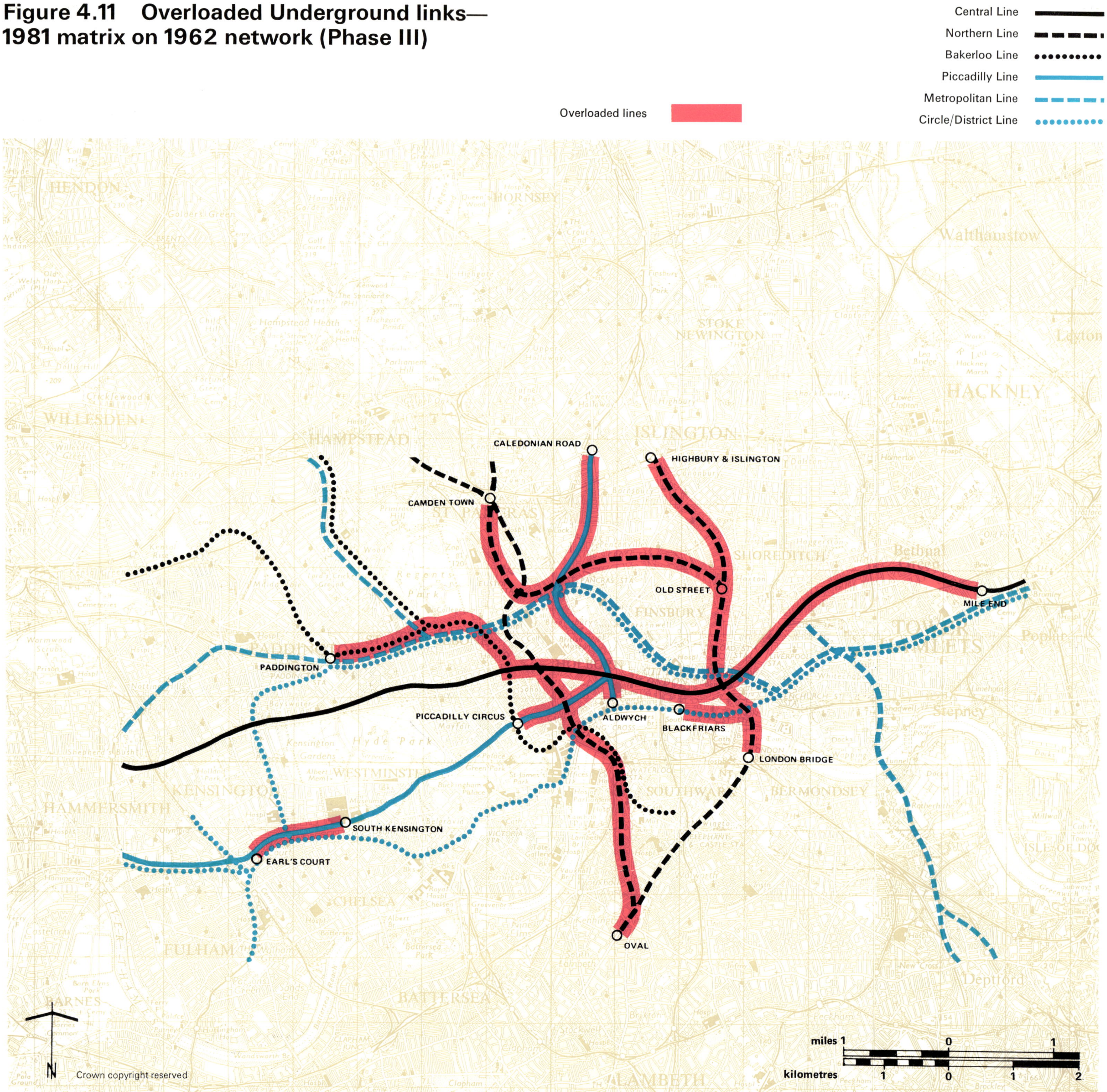

Figure 4.12
Overloaded Underground links—1981 matrix on plan C2/B/I (Phase III)

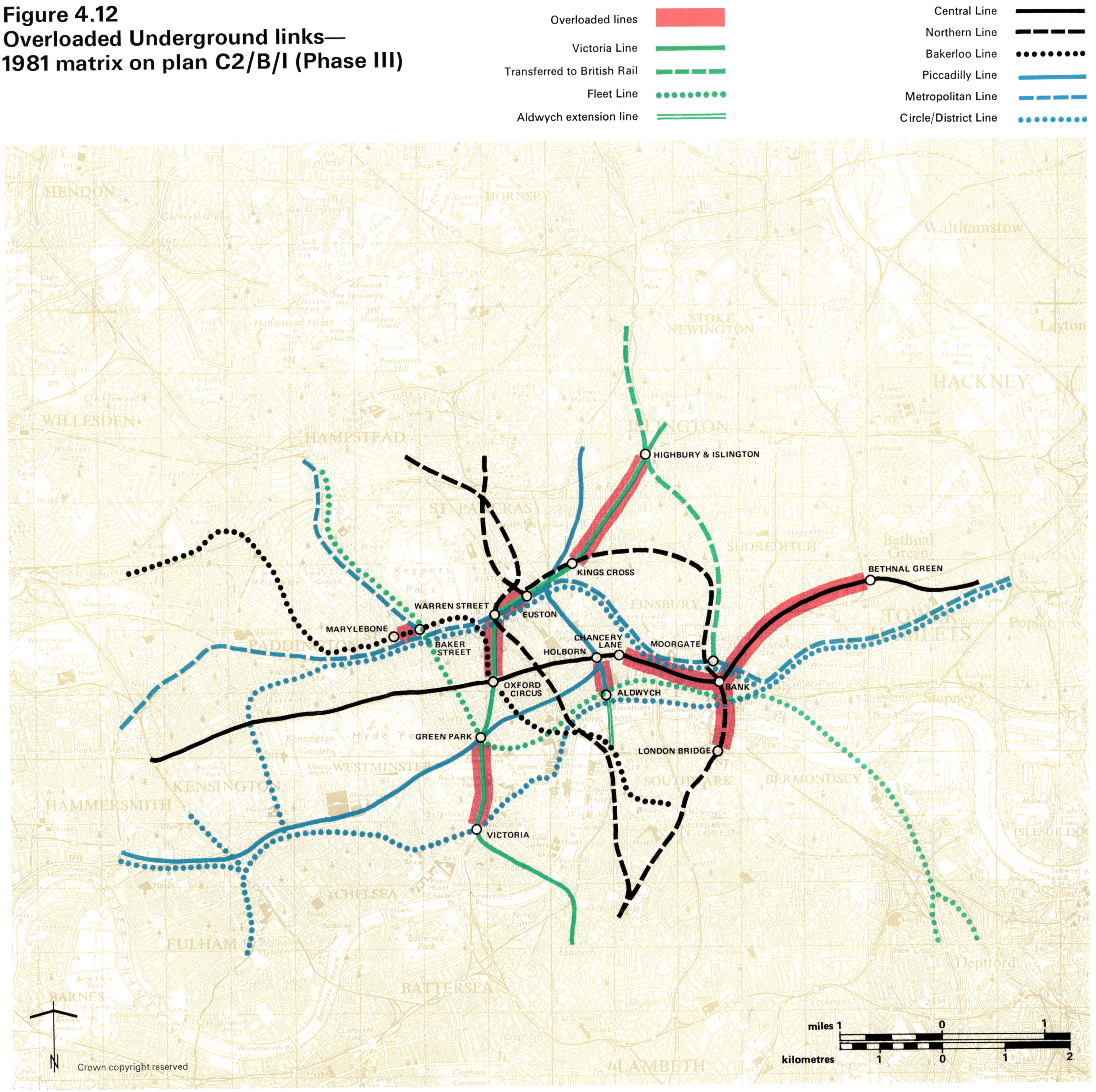

Design volume
as the volume used to select the number of lanes desirable in order that operating conditions do not fall below a chosen level.

Limiting volume
as the maximum volume which could be accommodated on a facility without undue risk of breakdown in flow at peak periods; it is somewhat less than the (absolute maximum) capacity of a facility.

4.50 The values for these definitions are shown in Table 4.2. On the basis that overloading is equivalent to an assigned flow equal to or in excess of the limiting volume and that underloading is equivalent to an assigned flow less than two-thirds of the limiting volume, the resulting over- and underloaded sections of plan 3 are shown in Figure 4.10. It should be emphasized that underloading still implies very large movements by current standards.

4.51 The precise location of these sections is less important than the general pattern which suggests that certain portions of Ringways 1 and 2 would be overloaded particularly to the north and west. Substantial parts of Ringway 3 to the east of London would be underloaded. Overloading does not necessarily imply a high priority for building since a comparison of benefits with costs must be made for this purpose. Nevertheless the very high degree of use of the inner motorways as opposed to the outer is noteworthy.

4.52 The interpretation of model output for the purposes of operational evaluation is not so well developed for public transport as for roads. Therefore only a limited evaluation was carried out on the Underground system in central London. Overloading on a link was defined as a passenger flow in excess of 80 per cent of the maximum number of sitting and standing places on a train multiplied by the number of trains per hour on that link. Figure 4.11 shows the overloading predicted by the model if no improvements were to be made beyond the 1962 network. Overloading on the Central Line extended as far as Oxford Circus from Mile End; that on the Piccadilly Line to Piccadilly from Holborn and Caledonian Road to Aldwych. On the Northern Line the overloading extended from Oval to Tottenham Court Road, Camden Town to Old Street and Highbury and Islington to London Bridge whilst on the Piccadilly and Circle/District Lines it occurred between Earls Court and South Kensington and between Blackfriars and Monument respectively. Finally the overloading on the Bakerloo Line occurred between Paddington and Oxford Circus. Figure 4.12 shows how this might be reduced by the introduction of plan C2/B/1 with, in particular, the Fleet Line and Victoria Line added. The only excessive overloading remaining would be between Marylebone and Baker Street on the Bakerloo Line, on the Central Line between Chancery Lane and Bethnal Green, on the Northern Line between Bank and London Bridge and on the Victoria Line between Euston and Oxford Circus, Green Park and Victoria and Highbury and Islington and Kings Cross.

Table 4.21 Proportion of trip purposes (Phases I and III)

	Private trips				
	Driver only		*All including passenger*	*Public transport trips*	
Purpose	*Phase I 1962*	*Phase III 1981*	*Phase III 1981*	*Phase I 1962*	*Phase III 1981*
Work	38%	29%	23%	57%	50%
Other home-based	41%	52%	59%	34%	39%
Non-home-based	21%	19%	18%	9%	11%
Total	100%	100%	100%	100%	100%

Economic evaluation

4.53 The cost estimates for the various plans have already been described, less precision being possible for the public transport plans. The results of the traffic analyses enable an estimate to be made of the benefits obtained when two plans are compared.

4.54 For comparisons between road plans the benefits have been estimated by trip type and trip purpose under three headings:

a. Benefit due to change in trip costs.

b. Benefit due to increased capacity of the system.

c. Benefit measured in terms of the tax paid by generated traffic.

These benefits include only those which can readily be quantified. They exclude for example the environmental benefit resulting from traffic being taken away from secondary roads and channelled on to motorways, while they also exclude any adverse effects of the motorways on their immediate vicinity. The procedures and unit costs used are outlined in Appendix F. Table 4.22 shows the incremental benefits of road plan 3 over road plan 1, and Table 4.23 shows the same information for road plan 9 over road plan 1; in each case the public transport plan is the same, namely plan C2/B.

Table 4.22 Incremental travel benefits* of plan 3 over plan I (Phase III)

	Weekday				*Weekend*				
Purpose	*A*	*B*	*C*	*Total*	*A*	*B*	*C*	*Total*	*Total*
Work	4·0	10·6	—	—	—	—	—	—	—
Other home-based	5·8	13·5	—	—	—	—	—	—	—
Non-home-based	3·3	7·7	—	—	—	—	—	—	—
Internal driver	13·1	31·8	22·6	67·5	6·0	13·9	10·8	30·7	98·2
Internal commercials	3·5	3·5	2·8	9·8	1·0	1·0	0·8	2·8	12·6
Total internal	16·6	35·3	25·4	77·3	7·0	14·9	11·6	33·5	110·8
External	4·5	2·6	3·3	10·4	1·7	1·0	1·4	4·1	14·5
Total	21·1	37·9	28·7	87·7	8·7	15·9	13·0	37·6	125·3

*Annual benefit in £ million in 1981 (at 1966 prices)
A—Benefit due to change in trip costs B—Benefit due to increased capacity of the system C—Benefit measured in terms of the tax paid by generated traffic

Table 4.23 Incremental travel benefits* of plan 9 over plan I (Phase III)

	Weekday				*Weekend*				
Purpose	*A*	*B*	*C*	*Total*	*A*	*B*	*C*	*Total*	*Total*
Work	6·8	12·3	—	—	—	—	—	—	—
Other home-based	10·1	16·2	—	—	—	—	—	—	—
Non-home-based	5·3	9·0	—	—	—	—	—	—	—
Internal driver	22·2	37·5	30·9	90·6	10·4	16·6	14·7	41·7	132·3
Internal commercials	5·5	3·5	2·7	11·7	1·6	1·0	0·8	3·4	15·1
Total internal	27·7	41·0	33·6	102·3	12·0	17·6	15·5	45·1	147·4
External	5·7	2·6	3·1	11·4	2·2	1·0	1·3	4·5	15·9
Total	33·4	43·6	36·7	113·7	14·2	18·6	16·8	49·6	163·3

*Annual benefit in £ million in 1981 (at 1966 prices)
A—Benefit due to change in trip costs B—Benefit due to increased capacity of the system C—Benefit measured in terms of the tax paid by generated traffic

4.55 The comparison of road plans is shown in Table 4.24. The rates of return calculated are on the basis of one-year incremental benefits (at 1966 prices) compared with total capital costs including parking provision. Thus no account has been taken of the time stream of benefits and costs by means of a discounting procedure. The rates of return for the road plans are based on the assumption in the traffic model that the system would operate at conditions of maximum flow, which implies that some means of control, both technically and politically feasible, can be developed to ensure that this is the case.

Table 4.24 Incremental benefits,* costs and rates of return for road plans (Phase III)

Plans	*3 v 1*	*9 v 1*
Annual travel benefit	125·4	163·2
Annual maintenance	—3·6	—4·6
Annual net benefit	121·8	158·6
Network costs	1286·0	1691·0
Parking costs	104·0	134·0
Total costs	1390·0	1825·0
One year rate of return	8·8%	8·7%

*Annual benefits and costs in £ million in 1981 (at 1966 prices)

4.56 The incremental rate of return of one plan over another is reduced if it is possible to operate the smaller network efficiently. This can be explained as follows: assume for a moment that by means of a restraint policy such as road pricing or direct controls it is possible to operate each plan in the manner assumed by the model and that,

B_1 = benefit of plan 1 operating under congestion

b_1 = additional benefit of restraint to plan 1

with similar definitions for plan 3. If the benefit from the restraint policy is greater for the smaller network than for the larger, as was the case with these analyses then

$b_1 > b_3$

Add $(B_3 - B_1)$ to both sides

$\therefore (B_3 - B_1) - (b_1 - b_3) < (B_3 - B_1)$

$\therefore (B_3 + b_3) - (B_1 + b_1) < (B_3 - B_1)$

which means that the incremental benefit with effective restraint policies is less than without such policies.

4.57 If no restraint were applied and network equilibrium relied on congestion, as happens today, the calculated additional benefit is as shown in Table 4.25. Thus the one-year rates of return vary between approximately 10 and 20 per cent depending on the way the networks would be operated. It is unlikely that traffic demand will be allowed to produce excessive congestion in future, but it is not clear to what extent control measures can or will be adopted. If control measures are adopted they are unlikely to be refined enough to operate at the point of maximum flow assumed in this analysis. It is likely therefore that the actual one-year rate of return on a complete motorway system in London lies between 10 and 20 per cent.

4.58 The economic assessment of public transport plans on a whole London basis presents even more difficult technical problems than that for the roads. The benefits that could be quantified in the study included:

a. Benefit due to passenger time saving.

b. Benefit due to reduced bus operating cost.

c. Disbenefit due to increased railway operating cost.

d. Benefit measured in terms of receipts from generated traffic.

An important omission is the benefit arising from allowing passengers to travel more comfortably with less overcrowding and more conveniently with fewer interchanges. The order of magnitude of the benefits that could be quantified is illustrated in Table 4.26.

4.59 Because of the uncertainties associated with the level of public transport demand on the central London links, the inexact nature of the cost estimates and the limited quantification

Table 4.25 Effect of control on rate of return (Phase III)

Plans	*Benefit**	*One year rate of return*
3v1		
With control	121·7	8·8%
Without control	296·8	21·4%
9v1		
With control	158·7	8·7%
Without control	373·6	20·5%

*Annual benefit in £ million in 1981 (at 1966 prices)

Table 4.26 Incremental travel benefits* of plan G/B over plan C2/B (Phase III)

Purpose	A Weekday	A Weekend	B	C	D	Total
Work	1·05	0·13	—	—	—	—
Other home-based	0·35	0·18	—	—	—	—
Non-home-based	0·74	0·22	—	—	—	—
Total internal	2·14	0·53	—	—	—	—
External	0·21	0·05	—	—	—	—
Total	2·35	0·58	2·54	−2·17	0·02	3·32

*Annual benefit in £ million in 1981 (at 1966 prices)

A—Benefit due to passenger time saving
B—Benefit due to reduced bus operating cost
C—Disbenefit due to increased railway operating cost
D—Benefit measured in terms of receipts from generated traffic

of benefits, a rate of return based on this analysis is not as representative as the rate of return for the analysis of the road networks.

Variations in population and employment

4.60 At the start of Phase III six relatively simple tests were carried out to test the sensitivity of road traffic volumes for internal driver trips to changes in population and employment distribution. They were:

Test 1

The 1981 employment in five outlying sub-centres, Kingston, Harrow, Edmonton, Ilford and Bromley, was doubled, this increase being offset by a corresponding decrease of 164,000 in Central Area employment.

Test 2

The previously forecast 1962–81 Central Area employment growth of 192,000 was distributed equally between eight outlying sub-centres: Dartford, Bromley, Kingston, Surbiton, Watford, Ilford, Romford and Uxbridge. Such an increase would about double the 1962 size of these sub-centres.

Test 3

Each of the five sub-centres in Test 1 was allocated employment from the Central Area which brought their 1981 employment about equal to that of Croydon in 1962 (79,000). The loss to the Central Area would reduce its 1981 forecast employment to a level 13 per cent less than that existing in 1962.

Test 4

This test implied an ambitious and positive programme of decentralisation from London and the South-East Region. The 1981 employment and population totals from the non-central LCC area were assumed to be reductions, by 526,000 and 1,120,000 respectively, of the 1962 totals.

Test 5

Employment and its complementary population within the LTS Area were assumed to grow, by 526,000 and 1,120,000 respectively, between 1962 and 1981. All of this growth was assumed to be confined to ten sub-centres on or near Ringway 3: Dartford, Orpington, Epsom, Walton-on-Thames, Hayes, Ruislip, Watford, Potters Bar, Enfield and Romford.

Test 6

This was a combination of Tests 4 and 5 representing a reduction of 526,000 jobs and 1,120,000 people from the 1962 level in the non-central LCC area, and a 1962–81 increase of the same amounts shared by the ten sub-centres.

The tests indicated that large reductions in Central Area employment would not significantly reduce the largest volumes predicted on the high class road network. Also percentage changes in overall traffic levels were of the same order but slightly less than changes in employment and population in the non-central LCC area. However these tests were of a simple nature and work is continuing on the detailed implications of changes in population and employment.

Conclusions

4.61 The results of Phases I (1962) and II (1981) highlighted several important points about the existing and future travel pattern in London.

a. The small proportion of external trips.

b. The predominance of rail for work trips to the Central Area.

c. The predominance of the car for work and even more for non-work trips outside the Central Area.

d. The continued predominance of rail for work trips to the Central Area in future.

e. The increasing use of the car for trips to the Central Area in future especially for non-work purposes.

f. The decline in the use of the bus in future particularly for trips outside the Central Area.

g. The very large increase in future in the use of the car for work purposes and even greater increase for non-work purposes for trips outside the Central Area.

4.62 These results, so far as they relate to the future, assumed that car ownership would increase very rapidly and that cars would be used by their owners with the same degree of freedom as in the 1960's. This led to road traffic loads in excess of the capacity of the very substantial network which it was assumed in the analysis would be provided.

4.63 Phase III modified the forecasts to produce loadings approximately in balance with the different networks tested. This had implications not only for the road system but for the public transport system also. Three road networks were examined with substantially different capital costs. Plan 9, with a large capital cost, required no restraint. Plan 3, with a lower, but still large capital cost required a small degree of restraint of road traffic if it was to operate satisfactorily. Plan 1, with a quite low capital cost, produced traffic estimates which implied a very large measure of traffic restraint.

4.64 The amount of capital which will be available for road investment up to the 1980's is likely to lie between the costs of plans 1 and 3. It is therefore safe to conclude that, whatever network is in fact provided, it will not be able to meet in full the potential demand for movement by road. It follows that whatever investment is made in the road system a substantial degree of control of movement will be advantageous. The means whereby this control would be exercised might be through parking control, road pricing or other method or some combination of different methods. The LTS does not give guidance on the appropriate method of control, which will depend on technical feasibility and political acceptability.

4.65 Even with quite large increases in road investment there will not be any increase in the freedom of an individual car owner to use his car. This is because rapidly increasing car ownership will lead to intensified competition for road space. Alternatively if investment in roads is curtailed then the degree of control of the use of the car either by direct means or through congestion will be severe.

4.66 The economic analysis in Phase III is not based on discounted benefits, but calculates a one-year rate of return for 1981 by comparing incremental benefits between plans with incremental cost. The analysis suggests that a high proportion of benefit from road investment in future will result from increased mobility in terms of allowing increased numbers of trips to take place, particularly by car. It also suggests that the one-year rate of return on investment, over Plan 1, will lie within the range 10 to 20 per cent, depending on the level of expenditure and the control of individual freedom to use the private car.

4.67 The Phase III analysis, based on rather higher forecasts of population and employment in Inner London than are now thought to be likely, suggests that the use of public transport will not fall below present levels even with major investment in new roads. This is based on the assumption that people who have cars available, but are unable to use them because of control measures, will be willing to use public transport to a substantial extent to make their desired trips.

4.68 In Appendix G the overall pattern of movement in London is shown for test 4 of LTS Phase III and for a test of the Council's London Roads Objective network, which is described in Chapter 5. The new techniques used in Phase III, and subsequently, resulted in a large decrease in the number of driver trips and an increase in the number of public transport trips as compared with the results of Phase II. Neither of the tests suggests however that the overall geographical pattern of movement is likely to be substantially different from that foreseen in Phase II.

5

Chapter 5 Work undertaken by the Greater London Council

Introduction

5.1 The London Transportation Study is complete. The techniques and principal results have been presented in earlier chapters and in appendices to this volume. But transportation planning is a continuing process. Even though the LTS was large and complex it was not expected that it would cover all aspects of the transport problems of London for all time.

5.2 Further work is being undertaken by the Council and will continue in future. This is not the place to describe all such work, but in this chapter that part of it is summarised which has most relevance to the LTS and the formulation of transport policies for London.

5.3 A number of separate studies are described, not all of which are directly related to each other, but which either extend or fill in gaps in the LTS. Since all the studies of future transport are crucially dependent on forecasts of car ownership, work has been carried out on this subject. The forecasts to 1981 are not changed to any large extent from those used previously but are now dependent on assumptions about the cost of cars and rates of depreciation and income available for car purchase rather than just on total income. Thus the forecasts may be varied in future as information on these factors is revised.

5.4 Much of the work in the LTS was based on 24-hour two-way movements on links in the transport network. It is clear that for many purposes it is necessary to forecast peak hour one-way flows. The Greater London Council has co-operated with the Ministry of Transport in developing a study of this topic and the progress to date is reported. Much of the value of the LTS has been in the opportunity it provided to apply basic data to a whole series of questions as they arose. The Council has recently announced a parking policy for London. The analyses of LTS data which were carried out prior to the development of the policy statement are described.

5.5 It is necessary in transportation studies to carry out many of the analyses mathematically and the size and complexity of the problems requires resort to computers. The human element must not be ignored however. Of crucial importance in the development of plans for transport investment and management is people's attitude to the facilities provided and particularly the way in which they behave where they have a choice of transport mode. Equally important is the effect of transport on the environment of the city. It is necessary to try to define the term which is often used rather vaguely in this respect, *quality of life*, and to learn to measure the benefits and disbenefits to the environment of different transport proposals. Progress to date with studies of attitudes and of the environmental effect of traffic is briefly described.

5.6 Finally a test is described of the Council's London Roads Objective network. The LTS tested three possible future road networks in order to detect significant differences in their performance. In so doing a number of new study techniques were developed. These have now been used to examine a specific network proposal and the results are reported. The test included a downward revision of the population and employment forecasts for 1981 as compared with those used in the LTS. More recent work has suggested that the forecasts should be even lower. The implications of these most recent forecasts are being examined but the results are not yet available to be reported in this volume. As transportation planning becomes more sophisticated it is possible not only to examine whole networks but also to assess the effect of investing in the links of the network in different orders. A procedure has been developed and applied to some of the projects in the Council's programme which has led to specific policy recommendations and this work is described.

5.7 The chapter closes with a brief indication of the scope of the continuing studies which are necessary for transport planning in London and are already under way.

Car ownership

5.8 The forecasting of future levels of car ownership is crucial to the planning of transport systems. The possession of a car affects both the use of the road system and the use of the public transport system and has a considerable number of side effects such as the parking demand at both ends of a trip and impact on the environment.

5.9 Levels of ownership on a national scale have been predicted by the Road Research Laboratory. These predictions were based on estimates of the ultimate saturation level of car ownership using a mathematical relationship to link present day values with this saturation level. Volume II of the LTS related car ownership to household income, but ignored such factors as cost of cars and rates of depreciation. Work done by the Greater London Council has attempted to carry these earlier trend studies further. The results for the UK are summarised in Figure 5.1.

5.10 It will be seen that a much higher saturation level of car ownership than that forecast by the RRL is predicted, based on data from the high income groups surveyed in the LTS, but that the difference between the RRL and the GLC forecasts in 1981 is not large. In the years following 1981 the difference becomes significant and so regular checks on existing and forecast car ownership levels will be necessary.

Figure 5.1 Car ownership forecasts for the UK

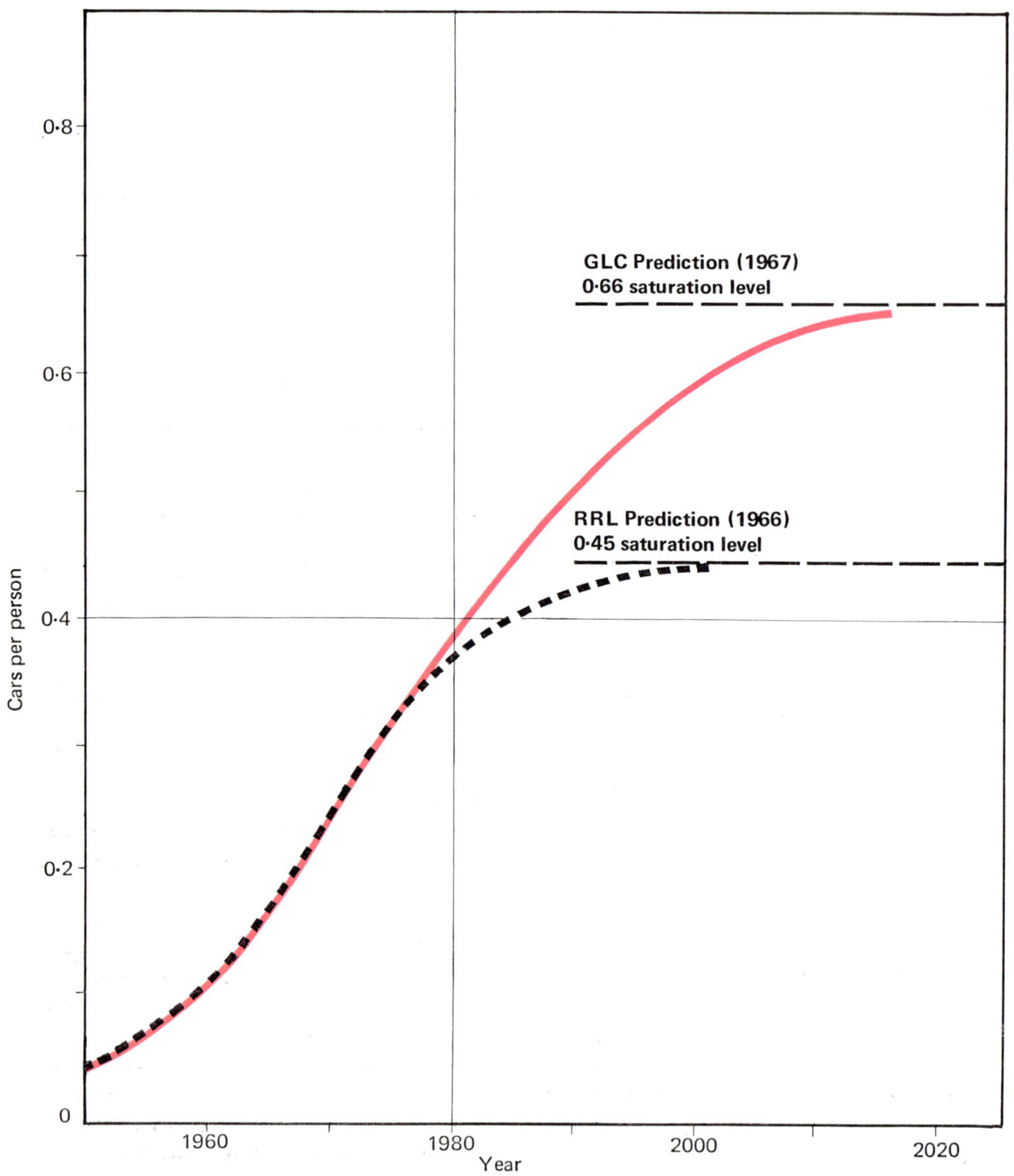

5.11 The same procedure which produced these results was used for the LTS area. Depreciation rates and the cost of cars bought by different income groups within the population were examined more closely. It was apparent that average car prices are falling in relation to income and that this fall is greatest for those cars being bought by the lower income groups. This has given rise to the concept of 'car purchasing income'. Car purchasing income can be defined as the disposable income of households relative to car prices. The price of new cars relative to real income is expected to fall slightly in the future but the price of second-hand cars is falling more rapidly the older they get. Thus the car purchasing income of income groups likely to buy second-hand cars is rising more rapidly than that of groups buying new cars. Figure 5.2 illustrates, for two predictions in income growth for the London area of 2·0 and 2·6 per cent per annum, the forecast real income and car purchasing income levels.

5.12 The car purchasing income distribution adjustment can be applied directly to the 1962 distribution of car ownership and the results related to real income to avoid having to make a new calculation of car purchasing income every time an alternative growth rate of income is examined. This method accepts implicitly particular assumptions about depreciation rates and the price of second-hand cars but is easier to apply. Table 5.1 shows calculations based on this method for income growths of 2·0 and 2·6 per cent. The car ownership calculations have been adjusted to take account of expected changes in residential density and the number of employed residents within the income groups used for the analysis. For the forecasts illustrated by Table 5.1 this correction was 2·3 per cent.

5.13 In Phases II and III, car ownership for 1981 was calculated as 68·1 per cent car-owning households, whilst in the analysis of the Council's London Roads Objective network referred to later, a figure of 66·8 per cent was used resulting from the application of the method described above.

5.14 These forecasts have been compared with the information available from the 1966 Census and registration records since 1962. The Census gives a low indication of total car ownership as it excludes cars not used exclusively by members of the household whether garaged at home or not. Vehicle registration records give a high estimate as they include vehicles normally garaged outside the GLC area. For 1966 the Census figures indicate that 42 per cent of the households are car-owning compared with 48 per cent from registration records. The mean of these two figures closely corresponds to an interpolation between 1961 and 1981 of the LTS area forecasts given above.

Peak hour factors

5.15 The LTS was based mainly on 24-hour estimates of traffic flow and capacity. For geometric design purposes model output is thus only available on a 24-hour basis. This has obvious limitations and to assist the interpretation of results a study of peak hour factors was commissioned by the Ministry of Transport from W S Atkins and Partners, in conjunction with work by the Greater London Council. The study was based on 1962 observed data and a relationship was defined:

$$Q_{ij} = \frac{\text{peak hour one-way flow from area i to area j}}{\text{24-hour two-way flow from area i to area j}}$$

Factors were derived for several trip purposes, for each peak (8–9 am,

Figure 5.2 Income distributions for the LTS area—1962 and 1981

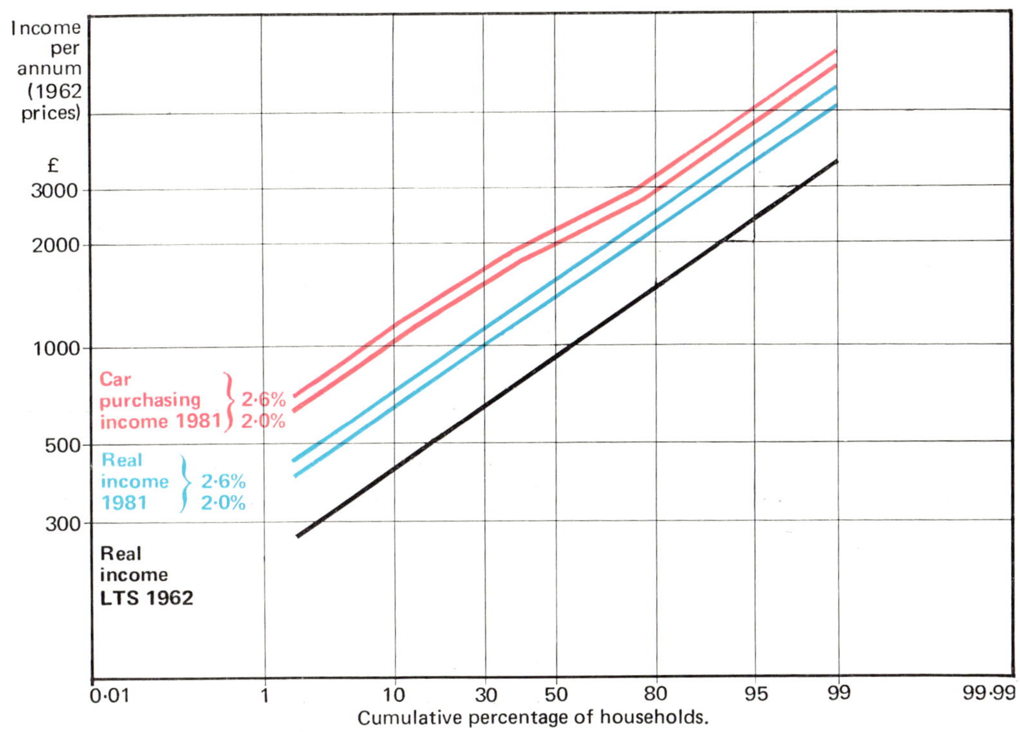

Table 5.1 Calculation of car ownership in the LTS area in 1981 for income growths of 2·0 and 2·6 per cent per annum from 1962

Range of total annual household income	*Percentage of households in range in 1981*		*Proportion with car*	*Percentage of households with car in 1981*	
	2·0	2·6		2·0	2·6
Under £500	5	3	0·13	0·7	0·4
£500–£999	26	21	0·46	12·0	9·7
£1000–£1499	27	26	0·69	18·6	17·9
£1500–£1999	18	19	0·80	14·4	15·2
£2000–£2499	10	12	0·85	8·5	10·2
£2500–£2999	6	8	0·90	5·4	7·2
£3000–£3999	5	6	0·95	4·8	5·7
£4000 and over	3	5	0·95	2·9	4·7
Total	100	100	—	67·3	71·0
Less 2·3 per cent adjustment				65·0	68·7

5–6 pm) for the LTS area divided into 53 areas, and for 15 external areas. Peak hour two-way factors by purpose were also derived and are shown for the whole area in Table 5.2.

5.16 The factors may be weighted by estimates of future numbers of trips for each purpose to calculate the overall peak hour factors for some future date and the steps are illustrated in Table 5.2. It was assumed that the peak two-way factors would remain the same in future for each trip purpose. They were equivalent in 1962 to overall factors for the area of 10·3 and 10·1 per cent for morning and evening respectively. The estimated proportions of trips by purpose in 1981 are shown in Table 5.2 and may be compared with the actual proportions in 1962. The estimated proportions were then used to weigh the peak hour factors by purpose which, when summed, gave estimated overall factors for the LTS area in 1981 of 9·9 and 9·4 per cent.

5.17 Thus there is very little change in the overall peak hour factors between 1962 and 1981. This is due to the fact that commercial and external trips were assumed to hold or increase their proportion of 24-hour trips and thus offset the fall in the work trip proportion. Moreover the two-way peak hour factors were assumed to remain constant for each purpose, no account being taken of the fact that congestion might well spread the peak for work and commercial trips.

5.18 This analysis may of course be used for any number of areas and for any assumptions about proportions of trips by purpose and changes in peak hour factors by purpose. Work on this topic is continuing.

Research for a parking policy

5.19 In order to develop rational car-parking standards for the GLC area it has been necessary to relate car parking provision to the capacity of the network which would carry vehicles to their destination. The essence of the required analysis was an examination

Table 5.2 Estimates of 1981 peak hour factors

Purpose	Percentage* 24-hour trips in 1962	Two-way peak hour factor by purpose		Percentage† 24-hour trips in 1981	Percentage in peak in 1981	
		8–9 am	5–6 pm		8–9 am	5–6 pm
Work	26·1	0·167	0·164	17·8	3·0	2·9
Other home-based	28·8	0·071	0·082	33·4	2·4	2·7
Non-home-based	14·8	0·037	0·101	12·5	0·5	1·3
Commercial and taxis	21·1	0·110	0·040	21·9	2·4	0·8
External	9·2	0·114	0·120	14·4	1·6	1·7
Total	100·0			100·0	9·9	9·4

*Observed
†Forecast in GLC test 7 (see paragraph 5·44)

of the balance between supply and demand estimates for parking in future. These estimates were intended not simply to be forecasts of the future situation, but to assess the implications of the continuation of particular patterns of behaviour and given car parking standards as well as assumptions about the road network which would be in existence at a given time in the future. A study of the likely balance or imbalance between supply and demand then allowed policies to be developed.

5.20 In carrying out the calculations which are described below, existing LCC standards were used. The principal ones were:

Dwelling houses
1 car space or garage per dwelling unit

Offices
1 car space per 2000 sq.ft.

Shops
1 car space per 2500 sq. ft.

5.21 The analyses were summarised by borough totals though the results presented here refer only to totals for the Central Area. The estimates of gross parking supply or number of car parking spaces is shown in Table 5.3. This shows that as a result of the Inner London Parking Area (ILPA), all free on-street parking space will be removed from the Central Area and will be replaced by a smaller number of metered on-street spaces. The 1981 gross off-street parking supply figures were obtained from the application of LCC parking standards to assumed rates of redevelopment of private supply and assumptions about the development of public supply. Since not all parking spaces are full at any one time and because it is desirable to have some spare capacity,

Table 5.3 Gross parking supply (number of spaces) in the central area

Year	On-street		Off-street			Total
	Free	Metered	Private residential	Private non-residential	Public	
1962	52,800	10,430	8,000	41,000	24,000	136,230
1966	40,400	15,180	13,000	47,000	27,000	142,580
1981	—	31,500	26,500	69,000	49,000	176,000

utilization factors were used to calculate the net supply. Maximum and minimum factors were derived, based on existing situations. It was assumed that the same factors would apply in the future. The calculation resulted in upper and lower values for net parking supply figures which are later compared with demand in Table 5.6.

5.22 The existing and forecast parking demand figures in the Central Area are shown in Table 5.4 for each of three categories of parking:

a. Residential parking.

b. Parking associated with the journey to work.

c. All other parking.

The parking demand was calculated from the base (overnight) demand and the hourly accumulation based on the arrival and departure pattern and 24-hour trip totals.

5.23 Two sets of demand figures were calculated for 1981. The first was taken from the results of the Phase III test of plan 3. In this test the traffic flow in 1981 in a 24-hour period into the Central Area was estimated to have increased by 50 per cent over the 1962 figure. For a number of reasons this figure may be too high. The complete motorway network will not have been built by 1981 and the increase in the capacity of the secondary road system into the Central Area is unlikely to keep pace with the capacity of the motorway system in any case. Therefore a second demand estimate was produced which assumed a 20 per cent increase in the 24-hour flow across the Central Area cordon between 1962 and 1981. It is this estimate which is shown in Tables 5.4, 5.5 and 5.6.

5.24 In Tables 5.5 and 5.6 a comparison is made between supply and demand in the Central Area. The policies which have been derived from analyses of these figures depend on an examination of the balance or imbalance between supply and demand. It should be borne in mind of course that balance

Table 5.4 Base and peak parking demand in the central area

	Base demand				*Peak demand*			
Year	*Residential*	*Work*	*Other*	*Total*	*Residential*	*Work*	*Other*	*Total*
1962	32,530	—	8,750	41,280	20,360	67,690	21,160	109,210
1966	38,300	—	9,280	47,580	24,000	68,810	22,870	115,680
1981	61,400	—	10,840	72,240	41,630	98,750	28,830	169,210

Table 5.5 Residential base demand and parking supply in the central area

				Public supply (gross)			*Percentage public supply for residential base demand*
Year	*Base demand*	*Net residential supply*	*Excess demand*	*On-street*	*Off-street*	*Total*	
1962	32,530	7,600	24,930	63,230	24,000	87,230	29
1966	38,300	12,300	26,000	55,580	27,000	82,580	31
1981	61,400	25,200	36,200	31,500	49,000	80,500	45

Table 5.6 Peak demand and parking supply in the central area

		Net supply		*Shortfall*	
Year	*Demand*	*Maximum*	*Minimum*	*Minimum*	*Maximum*
1962	109,210	118,180	107,300	−8,970	1,910
1966	115,680	122,150	110,250	−6,470	5,430
1981	169,210	147,000	131,340	22,210	37,870

between supply and demand over the large Central Area does not necessarily imply local balance. For example, the borough figures suggest that for that part of the Central Area to the south of the river there is always an excess of supply over demand even when the reverse is true over the whole area. Even in the south, however, there may be pockets where demand exceeds supply.

5.25 The comparison of residential base (overnight) demand with supply in the Central Area is shown in Table 5.5. In calculating the net supply figure it was assumed that there would be 95 per cent utilization of residential space overnight. Based on the assumptions about increasing demand and supply it is clear that even though a high proportional increase in supply can be expected an increasing percentage of overnight parking will be in public spaces, both on-street and off-street.

5.26 Parking supply and demand in the Central Area are compared for the peak in Table 5.6. The results suggest that in recent years supply and demand have been in balance. The estimates for 1981 however show a parking shortage compared with the estimated demand. If redevelopment took place at the rate assumed, with LCC parking standards there would be a shortfall of some 30,000 spaces. In these circumstances there are two possible alternatives; either increase the supply of parking to meet the estimated demand or selectively control the demand to whatever level and type of supply seems appropriate. It was on the basis of these considerations that the parking policy was developed.

Studies of modal choice

5.27 As part of the work of Phase III of the LTS, traffic district 90 located in Richmond was used for a case study to assess the attitude of residents to the use of their cars and public transport.

5.28 The study examined a wide range of transport problems resulting mainly in qualitative conclusions which, although applying specifically to Richmond, are indicative of attitudes to transport in general. It was found that for the large majority of people the availability of a car is the main factor which determines which mode is used for a vehicular trip. For certain purposes the trip will only be undertaken if a car is available. When a car is available other modes will only be used if there are perceived difficulties in the use of the car. It should be noted that ownership of a car does not necessarily imply availability of a car. A car is not available to a wife for shopping if it must be used for the husband's work. Alternatively a car is not available for commuting to work if it is essential for the wife in taking the children to school.

5.29 The study revealed that in Richmond rail public transport is principally regarded as a mass transport facility orientated towards the centre of London and its main competitive appeal is for trips in which relative speed is a major criterion and for which it offers the potential of direct travel. For shorter journeys and ones requiring changes its appeal is diminished. These results may not be generally applicable to London particularly in areas where the Underground system has a strong distribution function. In Richmond both British Rail and London Transport lines are directed towards the Central Area.

5.30 The study in Richmond also suggested that the bus is regarded as a mode for local trips and in general is not regarded as being in competition with other modes. The availability of a car will eliminate consideration of the bus alternative. On shorter journeys the bus faces competition from walking. The bus will be used by non-car-owners primarily and its competitive appeal is strengthened by service frequency on short distance journeys although reliability is also important and erratic waiting times cause considerable discontent.

5.31 A study was also made of the public's general awareness of congestion and their attitude towards any form of restraint. Generally congestion is perceived but the individual has difficulty in recognising his own contribution to it. Only in certain circumstances for longer distance journeys to Central London does it, when combined with parking difficulties, affect the decision to use a car. For inter-suburban journeys these factors seem to have no effect and younger people, who have grown up with traffic congestion, are less deterred by it than older people. In principle the necessity for restraint is accepted by car owners but they do not necessarily understand that it will affect themselves. It would be most likely to be accepted if it went hand in hand with a programme of road building and public transport improvement. Local restraint was visualised in terms of reserving local facilities for local people although how it could be achieved was not generally understood.

5.32 Following the completion of the LTS it was decided to continue the studies of modal choice with the object of attempting to quantify a number of the qualitative impressions gained. A study on a much larger scale was therefore commissioned from Research Projects Ltd, who had carried out the Richmond Study.

5.33 The study was subdivided into a number of smaller studies. The main studies were based on a Central Area and a non-Central Area location with the object of determining the main factors in selecting the mode of journey to work. Other studies examined the use made of competing modes for a given journey along common routes, the use made of an interchange, the effect of the frequency of services serving an interchange on its use and the effect of distance from the rail system on the use of that system.

5.34 The general conclusions that can be drawn from this study have considerably helped the understanding of the extremely complex situation that exists in London. Some quantification of modal choice factors has been possible. The principal conclusions are as follows:

a. For journeys other than to Central London the availability of a car is the

rates in 1981. It was estimated that, if traffic levels were to remain the same, the overall number of accidents would remain constant but that increased traffic would result in more accidents. It was estimated that delays to pedestrians would show a marked increase on the one or two streets likely to have considerable increases in traffic flows.

b. *Noise*—Traffic noise measurements in the zone were used to draw a decibel contour map showing noise levels for day and night periods throughout the area. A similar map was drawn for 1981. It showed that the removal of most of the through traffic from the zone would slightly reduce levels generally and in no areas worsen them. Existing levels generally are considerably higher than those recommended by the Wilson Committee, especially around schools.

c. *Visual intrusion*—Attempts were made to assess the visual impact of both moving and stationary motor vehicles but the results to date have been subjective and inconclusive. Unfortunately it appears that the streets least able to cope with such visual intrusion are those where the greatest accumulation of parking will occur.

5.42 The study is not complete but certain conclusions are already apparent:

a. Taking the area as a whole the forecast level of car ownership could be accommodated in off-street space and by extensive use of kerbside space. However in the higher density parts of the area car owners will have to park some distance away from their home if present levels of service on the local roads are to be maintained in future. Local pressures may lead to the provision of more off-street space or possibly to reductions in the level of service on local roads. If this does not happen the inconvenience of parking may limit the growth of car ownership but the effect, if it happens at all, will be in particular small areas. The available evidence therefore suggests that the general magnitude of the 1981 car ownership forecasts is not likely to be diminished by physical constraints on parking in residential areas.

b. The effect of Ringway 2 and the improved A3, despite the large increase in traffic which is expected in London as a whole, would be to maintain the total level of traffic within the zone at approximately the present level. The proportion of this traffic which would be local is however much higher. The general inference is that when neighbouring motorways remove the majority of through traffic from an area the roads within that area can be adapted primarily to meet local needs and requirements.

c. Although noise levels in 1981 are not expected to meet Wilson Committee recommendations, they are expected to be, in general, a little better than those at present if through traffic is able to be channelled on to new and improved major roads.

d. If no new major routes are built which remove through traffic from the area then traffic flows are likely to increase with consequent increases in accident and noise levels.

5.43 The next stage of the study is in progress and is concerned with further work on visual intrusion, noise, vibration and fumes and with design schemes having the object of improving the environment by modifications to buildings. It is hoped that a similar study may be carried out in a zone somewhat further from the primary route system.

An examination of the proposed road network (GLC test 7)

Introduction

5.44 Phases II and III of the LTS examined alternative strategic transportation networks for London. In 1967 the Greater London Council, in consultation with the Ministry of Transport, produced a specific primary road plan for discussion and action which was thought to be attainable by the mid 1980's. It was published in 'London's roads—a programme for action' and was similar in overall pattern to plan 3 tested in Phase III of the LTS but included fewer links with some of them half the width of those in plan 3. The plan was formed after careful consideration of the traffic information and forecasts and of the available opportunities to build major new highways in London and was estimated to cost £650 million, with an expenditure of £210 million on secondary roads in addition. It was important that this network should be carefully examined to see how it would behave under the likely demand in the 1980's. A comprehensive test was carried out by the Research Group of the GLC Department of Highways and Transportation using the procedures developed during Phase III. The opportunity was taken to incorporate revised estimates of future planning parameters and to make some minor improvements to the model. To distinguish it from tests carried out in the LTS the work is referred to here as GLC test 7.

The networks examined

5.45 The road network examined in this test was based on the network published in 'London's roads—a programme for action' but included only that portion of the network which was then judged to be capable of completion by 1981. The excluded routes were the south part of Ringway I and its connection to the M.23. The secondary system was identical to that used in the LTS tests. The number and location of interchanges between secondary roads and motorways were appreciably different from plan 3 in Phase III, resulting in a much higher frequency of interchanges on Ringways 2 and 3 in particular. This was due both to the realignment of these roads and to the steady development of the network as planning proceeded within the Council.

5.46 Figure 5.4 shows the form of the network tested. It may be compared with Figure 4.3 which shows plan 3. Table 5.7 shows the actual mileage by location of the network and compares it with plan 3 in Phase III. 70 per cent of the plan 3 mileage is included and 53 per cent of the capacity. Table 5.8 shows how much more closely spaced are the interchanges on this network than on plan 3, particularly on Ringway 3. The public transport network used in the test was C2/B/1 which is described in paragraph 4.23.

Figure 5.4 Network for GLC test 7

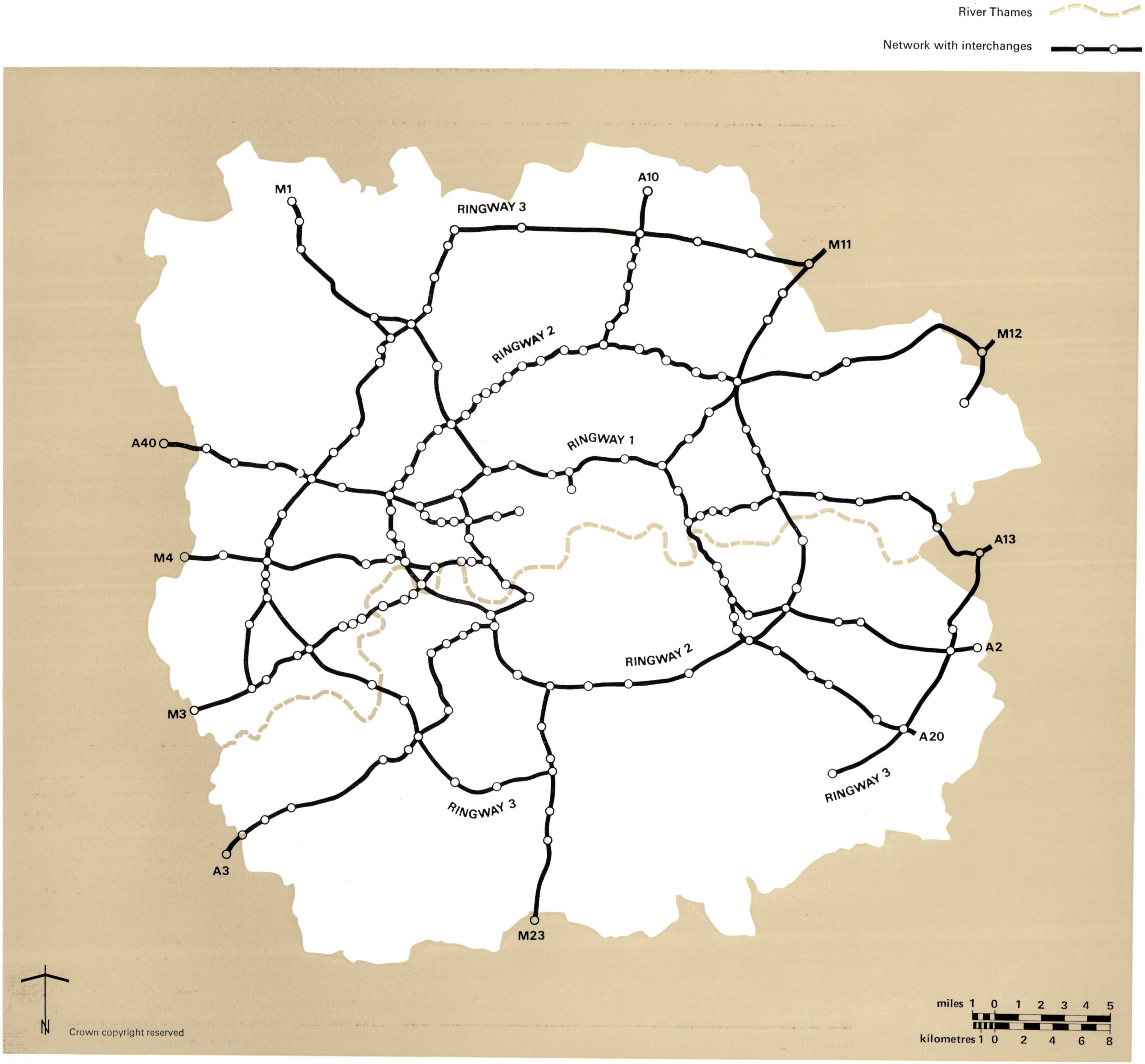

Table 5.7 Mileage of motorways in test network (GLC test 7)

Number of equivalent motorway lanes	2×1½	2×2	2×2½	2×3	2×4	*Total*	*Percentage of plan 3*
Ringway 1	—	6·5	—	4·8	10·4	21·7	73%
Ringway 2	—	25·6	21·2	—	2·1	48·9	100%
Ringway 3	—	15·2	4·4	36·4	6·3	62·3	62%
Inner radials (Ringway 1 to Ringway 2)	8·0	1·5	2·9	11·1	1·5	25·0	57%
Outer radials (Ringway 2 to Ringway 3)	21·9	1·2	17·9	31·8	4·4	77·2	68%
Inside Ringway 1	—	0·8	—	1·9	—	2·7	—
Beyond Ringway 3	—	8·3	11·0	29·5	—	48·8	—
Total	29·9	59·1	57·4	115·5	24·7	286·6	70%*

*Excludes mileage inside Ringway 1 and beyond Ringway 3

Table 5.8 Average spacing* of motorway interchanges

	GLC test 7 network		*Plan 3*	
Location	*All interchanges*	*Local access points*	*All interchanges*	*Local access points*
Ringway 1	1·4	1·9	1·3	1·7
Ringway 2	1·3	1·6	1·5	2·3
Ringway 3	1·9	2·4	2·9	5·5
Radials	1·4	1·8	1·5	1·9

*Spacing in miles

Revised planning forecasts

5.47 Revised forecasts of population and employment were used in this test. From Table 5.9 it will be seen that there are some significant changes in population estimates from the LTS figures, particularly in sectors 4 and 7. In general both population and employment figures are lower but some sectors do show increases, Appendix E makes comparisons with the 1962 situation. There are now available even later estimates based on the totals for the GLC area in the Greater London Development Plan. They were not available at the time of test 7. They are included, however, in the most recent work being undertaken but this is not yet complete.

Table 5.9 Planning parameters (GLC test 7)

Sector	*Population in households**		*Population employed**		*Total employment**		*Retail, general and other employment**	
	Test 7	*Percentage change over LTS*	*Test 7*	*Percentage change over LTS*	*Test 7*	*Percentage change over LTS*	*Test 7*	*Percentage change over LTS*
0	1·34	+0·4	0·67	−4·3	0·66	−10·8	0·36	−7·2
1	0·97	−0·8	0·47	−2·5	0·36	−17·1	0·19	−20·6
2	0·27	+1·1	0·13	+7·1	0·13	+8·7	0·06	−7·4
3	1·20	+1·2	0·59	−2·9	0·49	−3·9	0·25	−4·2
4	1·25	−10·1	0·66	−12·8	0·64	−2·7	0·38	−7·8
5	0·20	+0·5	0·11	+4·7	1·38	−8·8	0·98	−9·5
6	0·03	0	0·02	−5·9	0·12	+3·3	0·08	+14·1
7	1·50	−3·8	0·75	−5·6	0·52	+0·8	0·32	0
8	0·65	+4·5	0·31	+7·9	0·21	+10·1	0·11	+8·5
9	1·23	−1·8	0·58	−1·2	0·50	−2·9	0·29	−8·0
Total	8·64	−2·0	4·30	−3·8	5·01	−5·7	3·02	−7·2

*Millions of persons

5.48 Revised and lower forecasts of income growth were made which anticipated rates of growth of disposable income of between 2·0 and 2·6 per cent per annum in real terms, depending upon assumptions about growth of the GNP. It was decided to adopt a mean value of 2·3 per cent per annum, which compares with 2·8 per cent used in the LTS.

5.49 Revised forecasts of car ownership were also made, based in part on the revised income forecasts, and resulted in an overall figure of 66·8 per cent of car-owning households as compared with 68·1 per cent in the LTS, as described in paragraph 5.13. The reason for the small change in car ownership in spite of the significant fall in the assumed income growth is because this fall was offset by the assumed increase in car purchasing income (i.e. a fall in car prices 1962–1981) giving an overall increase similar to that used in earlier LTS work.

Revised trip ends

5.50 1981 trip generations were recalculated using the new planning parameters. A simplified procedure was adopted in which the LTS generations for each trip purpose (work, other home-based, non-home-based) were multiplied by a factor for each district in order to avoid having to calculate generations for each of the 81 categories used in the LTS. With the attraction end of trips a more basic approach had to be adopted. Because of revised estimates, particularly of employment for 1962, new regression equations had to be derived and then applied to the forecast year. The commercial and external trip ends and matrices were used unchanged from Phase III.

5.51 Table 5.10 lists the input trip ends for this test, compared with Phase III test 4 (post-restraint). The differences between the two forecasts result from the differences in the population, employment and income forecasts.

Table 5.10 Input trip ends*

Class and purpose of trip	*Phase III 1981*	*GLC test 7 1981*
Car-owners (excluding passengers)		
Work	3·98	3·61
Other home-based	5·56	5·29
Non-home-based	1·56	1·48
Total internal car owners	11·10	10·38
Non-car-owners		
Work	1·34	1·22
Other home-based	0·85	0·84
Non-home-based	0·27	0·26
Total internal non-car-owners	2·46	2·32
Unreported	0·82	0·82
Commercial and taxis	1·94	1·94
External—private and commercial	1·22	1·22
—public transport	0·85	0·85
Total	18·39	17·53

*24-hour average weekday millions of trip ends

Results

5.52 The numbers of road and public transport trips at the end of the test are shown in Table 5.11 where they are compared with the Phase III results. In comparison with plan 3 there are fewer road trips, offset by a small increase in public transport trips. This is in part due to the lower income forecast which leads to fewer car-owning households but more especially to the lower capacity of the network. Table 5.12 gives the results in greater detail and may be compared with the Phase III results in Table 4.10.

Table 5.11 Comparison of trip* totals

	Phase III 1981	*GLC test 7 1981*
Private and commercial driver	10·48	8·69
Public transport	7·53	7·73
Total	18·01	16·42

*24-hour average weekday millions of trips

5.53 The number of internal car driver trips restrained is three times as many as for plan 3 in Phase III. Tables 5.13 and 5.14 show the vehicle-miles and vehicle-hours for this network and the test of plan 3. There are considerable reductions in both the vehicle-hours and -miles as compared with the Phase III result and a considerable proportional shift to the secondary network on account of the smaller primary capacity. The absolute values of secondary road vehicle mileage and hours are not increased however.

5.54 Table 5.15 shows the sector to sector internal vehicle movements before and after restraint and Table 5.16 gives the sector to sector 24-hour public transport passenger movements. Comparison with Tables 4.15 and 4.17 shows the effect of the new planning parameters and reduced network on the results of the test of plan 3 in Phase III.

Evaluation of the network

5.55 Figure 5.5 shows the final overloaded and underloaded links of the primary road network on a 24-hour basis. These links generally correspond with the overloaded and underloaded links of plan 3 (Figure 4.10). For this network some links are undoubtedly overloaded now because they have a lower capacity than in plan 3. The south side of Ringway 2 is more heavily loaded because of the absence of the south side of Ringway 1. Ringway 3 in the south-west is now heavily loaded but has been brought closer to areas of high traffic generation. The underloading of much of the remainder of Ringway 3 is again noticeable.

5.56 No formal economic evaluation of the whole network has been carried out using the techniques used in Phase III. This is because no other network has yet been tested for comparison using the same planning forecasts. However it may be surmised that, using the same method of calculation, the one-year rate of return on this network would be of the same order as plan 3, compared with plan 1.

5.57 This estimate stems from the fact that the cost of the network, which is £650 million, is between that of plan 1

Table 5.12 Summary of trips* (GLC test 7)

	Private and commercial driver		*Public transport passenger*				
	Pre-restraint	*Post-restraint*	*NCO households*	*Mode choice*	*Restraint*	*Unreported*	*Total*
Work	2·06	1·54	1·22	1·55	0·38	0·21	3·36
Other home-based	3·82	2·90	0·84	1·47	0·25	0·19	2·75
Non-home-based	1·45	1·09	0·26	0·03	0·06	0·42	0·77
Internal driver	7·33	5·53					
Commercial and taxis	1·94	1·94					
Total internal	9·27	7·47	2·32	3·05	0·69	0·82	6·88
External	1·22	1·22					0·85
Total	10·49	8·69					7·73

*24-hour average weekday millions of trips

Table 5.13 Vehicle-miles summary for private and commercial trips*

Test	*Restraint*	*Trips*	*Vehicle miles* *Primary*	*Secondary*	*Total*
Phase III	Pre-	8·19	33·37	27·58	60·95
	Post-	7·81	31·81	26·30	58·11
GLC test 7	Pre-	7·72	23·43	30·78	54·21
	Post-	6·46	20·16	26·12	46·28

*24-hour average weekday millions of vehicle trips and miles

Table 5.14 Vehicle-hours summary for private and commercial trips*

Test	*Restraint*	*Trips*	*Vehicle hours* *Primary*	*Secondary*	*Total*
Phase III	Pre-	8·19	0·83	1·23	2·06
	Post-	7·81	0·79	1·18	1·97
GLC test 7	Pre-	7·72	0·68	1·36	2·04
	Post-	6·46	0·58	1·14	1·72

*24-hour average weekday millions of vehicle trips and hours

Table 5.15 Total internal vehicle sector to sector trips* (GLC test 7)

Generation sectors	Attraction sectors 0	1	2	3	4	5	6	7	8	9	*Totals*
0	1,148,460	68,995	30,624	14,377	130,497	85,946	4,497	32,718	4,445	82,247	1,602,806
	855,387	53,693	24,556	11,520	98,143	62,997	3,391	27,385	4,169	67,957	1,209,198
1	65,821	698,305	16,841	46,711	125,248	81,180	4,149	16,447	3,442	7,534	1,065,678
	50,475	564,943	13,451	39,037	93,359	58,993	3,091	14,092	3,364	6,147	846,952
2	37,168	23,560	205,257	2,615	10,544	12,170	626	3,439	480	1,938	297,797
	28,096	18,079	157,320	2,172	7,955	8,890	482	2,958	448	1,614	228,014
3	14,385	48,378	2,132	882,054	112,443	46,833	3,305	31,431	25,808	7,030	1,173,799
	11,162	42,134	1,787	677,947	84,304	34,848	2,581	27,185	25,216	5,717	912,881
4	70,298	61,630	4,764	52,308	605,211	206,092	9,987	58,841	6,299	18,716	1,094,146
	56,890	53,489	4,048	43,566	470,023	161,473	7,957	51,159	6,150	16,443	871,198
5	27,749	19,669	2,063	15,024	144,063	358,364	24,517	46,880	4,637	10,910	653,876
	22,364	16,708	1,784	12,705	120,172	315,606	21,833	42,174	4,480	9,703	567,529
6	1,764	1,072	129	1,439	7,410	24,855	10,418	15,305	838	1,426	64,656
	1,502	965	123	1,291	6,251	22,100	9,338	14,207	813	1,314	57,904
7	31,016	8,651	2,148	33,160	79,673	91,994	19,544	715,630	61,220	79,763	1,122,799
	24,906	7,692	1,806	25,982	62,080	70,015	15,662	592,398	58,141	64,693	923,375
8	6,543	4,036	601	32,544	16,098	36,534	3,414	110,108	494,308	32,420	736,606
	5,146	3,432	490	23,932	12,337	26,380	2,525	89,924	483,085	28,197	675,448
9	84,597	7,300	3,119	11,944	55,736	81,970	5,911	138,365	25,163	1,038,144	1,452,249
	64,030	6,031	2,483	9,200	41,603	59,209	4,387	109,287	23,169	855,568	1,174,967
Totals	1,487,801	941,596	267,678	1,092,176	1,286,923	1,025,938	86,368	1,169,164	626,640	1,280,128	9,264,412
	1,119,958	767,166	207,848	847,352	996,227	820,511	71,247	970,769	609,035	1,057,353	7,467,466

*24-hour average weekday internal vehicle trips: upper figure—pre-restraint; lower figure—post-restraint

Table 5.16 Total public transport passenger sector to sector trips* (GLC test 7)

	Attraction sectors											
Generation sectors	0	1	2	3	4	5	6	7	8	9	*External*	*Totals*
0	427,304	31,599	12,135	15,204	89,378	224,057	13,609	41,245	4,075	46,638	5,334	910,578
1	49,253	268,365	10,094	22,233	76,940	196,012	12,502	26,962	1,983	20,313	3,675	688,332
2	26,021	13,280	66,755	4,458	12,638	30,596	1,564	4,240	682	2,634	1,447	164,315
3	31,351	35,746	10,823	388,647	98,426	220,221	14,418	41,401	17,418	15,733	6,800	880,984
4	93,275	52,547	16,366	49,313	359,289	449,442	31,845	52,941	35,939	44,965	4,621	1,190,513
5	40,730	18,622	5,971	20,741	86,070	309,533	18,495	30,236	16,634	33,572	2,276	582,880
6	3,163	1,470	722	1,697	4,768	21,459	3,562	5,197	2,182	5,832	1,100	51,152
7	61,079	36,554	6,571	76,170	100,440	348,408	38,402	396,715	29,889	79,523	9,358	1,183,109
8	14,425	4,667	1,970	52,431	22,353	126,153	9,707	45,926	129,964	16,713	1,531	425,840
9	43,834	10,829	5,866	13,951	54,314	235,377	19,072	69,293	10,513	372,601	5,822	841,472
External	10,879	8,548	4,215	24,099	56,383	627,141	39,644	17,129	3,833	15,485	0	807,356
Totals	801,314	482,227	141,458	668,944	960,999	2,788,399	202,820	731,285	253,112	654,009	41,964	7,726,531

*24-hour average weekday person trips

and plan 3. Provided that they are all good road investments and that there are diminishing returns once a substantial network is built, then the one-year rate of return on the network tested to examine the Council's road proposal would be at least as great as the range of approximately 10 to 20 per cent calculated for plan 3 over plan 1.

Conclusions

5.58 The conclusions from the first test of the Council's proposed primary road network are similar to those drawn in Chapter 4 from a consideration the LTS Phase III results. These are of compared in Appendix G.

5.59 The network will not meet in full the potential demand for movement by road in the 1980's. There will need to be substantial restraint of movement by road of those with a car who would like to use it. The degree of restraint required is reduced however because of the lower population and employment estimates and particularly the lower income estimates used in this analysis.

5.60 The results do not show an increase in the amount of travel on the secondary road network, as compared with the test of the full plan 3 in Phase III, but it represents a greater proportion of the total road movement. In spite of the substantial restraint of road traffic the increase in number of public transport trips as compared with the test of plan 3 in Phase III is small. This level of public transport trip making, which is a slight increase on the 1962 level, would depend, as discussed in Chapter 4, on car owners who were restrained from using their car being willing to use public transport as an alternative.

Priorities

5.61 In order that detailed planning of the primary road network could proceed it was necessary to decide what the order of priority of schemes should be. To build the cheapest schemes first is not necessarily to serve the greatest need. Conversely, to build those schemes first which carry the greatest traffic load is not necessarily the best economic proposition. An analysis was carried out to ascertain whether there was a significant difference in the benefits to road users from different sequences of construction

Figure 5.5 Overloaded and underloaded links (GLC test 7)

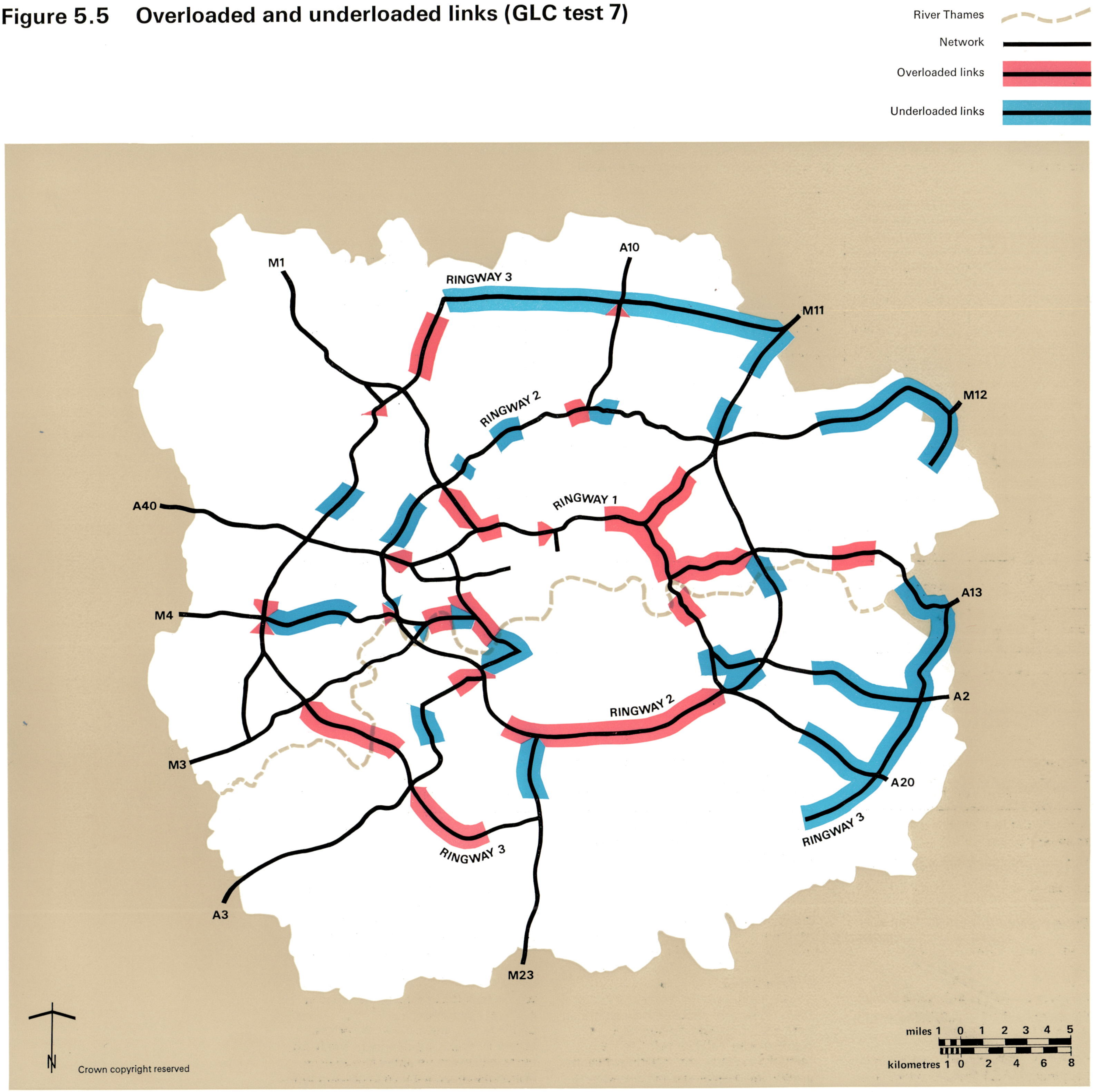

and, if so, what was the most desirable sequence.

5.62 A group of eight projects was defined for possible construction during the period 1971/72 to 1978/79. They would be built following the completion of the East Cross Route and of the West Cross Route to the river. It was also assumed that various trunk road schemes would be completed during this period—Ringway 3 between M.3 and A.40 and between A.1 and M.11, also the M.23, M.11 and M.12 to Ringway 2.

5.63 The eight projects considered were:

Project A

Ringway 2—A.13 to Thamesmead Spine

Ringway 2—Thamesmead Spine to Falconwood

Project B

Ringway 2—Falconwood to Sidcup Road link

Ringway 2—Sidcup Road link to A.21

A.20 improvement (metropolitan road portion)

Sidcup Road link—Kidbrooke to A.20

Project C

North Cross Route—Hackney Wick to York Way

Project D

North Cross Route—M.1 to A.40 link

West Cross Route—A.40 link to A.40

Project E

M.1 extension to Finchley Road

Project F

Eastern Avenue/M.11 extension—Hackney Wick to Ringway 2

Project G

Ringway 2—A.3, Wandsworth to M.4

Ringway 2—A.24 to A.3, Wandsworth

Project H

Ringway 2—A.21 to A.24

and they are shown in Figure 5.6 in relation to other existing or committed projects.

5.64 In a problem of this nature there are always political, financial and engineering constraints which affect the decisions which can be made. Nevertheless there is still considerable freedom which justifies a careful analysis of the choices available. This is because of the large number of possible sequences of construction.

5.65 A method was devised which reduced the size of the combinatorial problem of finding the optimum sequence of construction. Benefits were measured in terms of total travel time savings throughout the construction period. The time beyond this period could be ignored because the benefits accruing would be identical regardless of the sequence. The total construction cost was also taken as independent of sequence.

5.66 The benefits were then translated into monetary terms and discounted at 8 per cent to the year 1971/72 (at 1966 prices). The difference between the discounted totals was such that one possible sequence chosen without numerical analysis gave a discounted monetary benefit of £40 million as compared with £60 million for the optimum sequence. Thus the importance of the sequence of construction was demonstrated. The optimum sequence was H, G, A, B, E, D, F, C. Since the position of adjacent projects in the sequence is not considered to be of major significance, the results may be expressed in general terms as recommending the following order:

Ringway 2, south

Ringway 1, north-west corner

Ringway 1, north-east corner

Future work

5.67 Transport planning is a continuing process. It must be flexible and responsive to the most up to date information about existing travel patterns, people's personal and travel characteristics and attitudes to transport and the most recent estimates of future population and employment movements. As understanding is improved and as techniques develop it will be increasingly possible to integrate transport planning into general planning. Studies and plans must be responsive to the estimates of finance available for investment and, indeed, the studies must assess the implications of investing different total sums of money in transport and different distributions of such sums.

5.68 To these ends the Council is continuing its studies of transport. The assessment of the implications of new lower population and employment estimates has already been mentioned. Studies are being undertaken of the whole of the secondary road system in London, in terms of its capacity and cost of improvement to compare with estimates of demand for movement. The Council in cooperation with British Rail and London Transport are developing the methods of analysis of public transport which were introduced in the LTS so that improved forecasts for public transport on a comprehensive basis may be obtained and so that public transport analysis may be put on a comparable basis to that of private road transport. Methods of providing detailed traffic design data in a form most directly useful to the highway engineer are being developed and work will proceed in attempting to produce similar data for public transport.

5.69 In addition to the continued examination of the strategic aspects of transport the study of individual major projects will be extended both in terms of traffic and economic analysis of schemes and the study of feasible alternatives where they exist. This will require further development of methods of cost-benefit analysis.

5.70 Meanwhile the existing travel situation will be regularly monitored. A programme of traffic counts and road speed measurements is under way and will be reported at regular intervals. In addition a programme is being prepared for the regular updating of the Traffic Survey so that a check may be made on the relationships obtained in 1962 between person and travel characteristics.

Figure 5.6 Priority assessment projects on plan 3 background

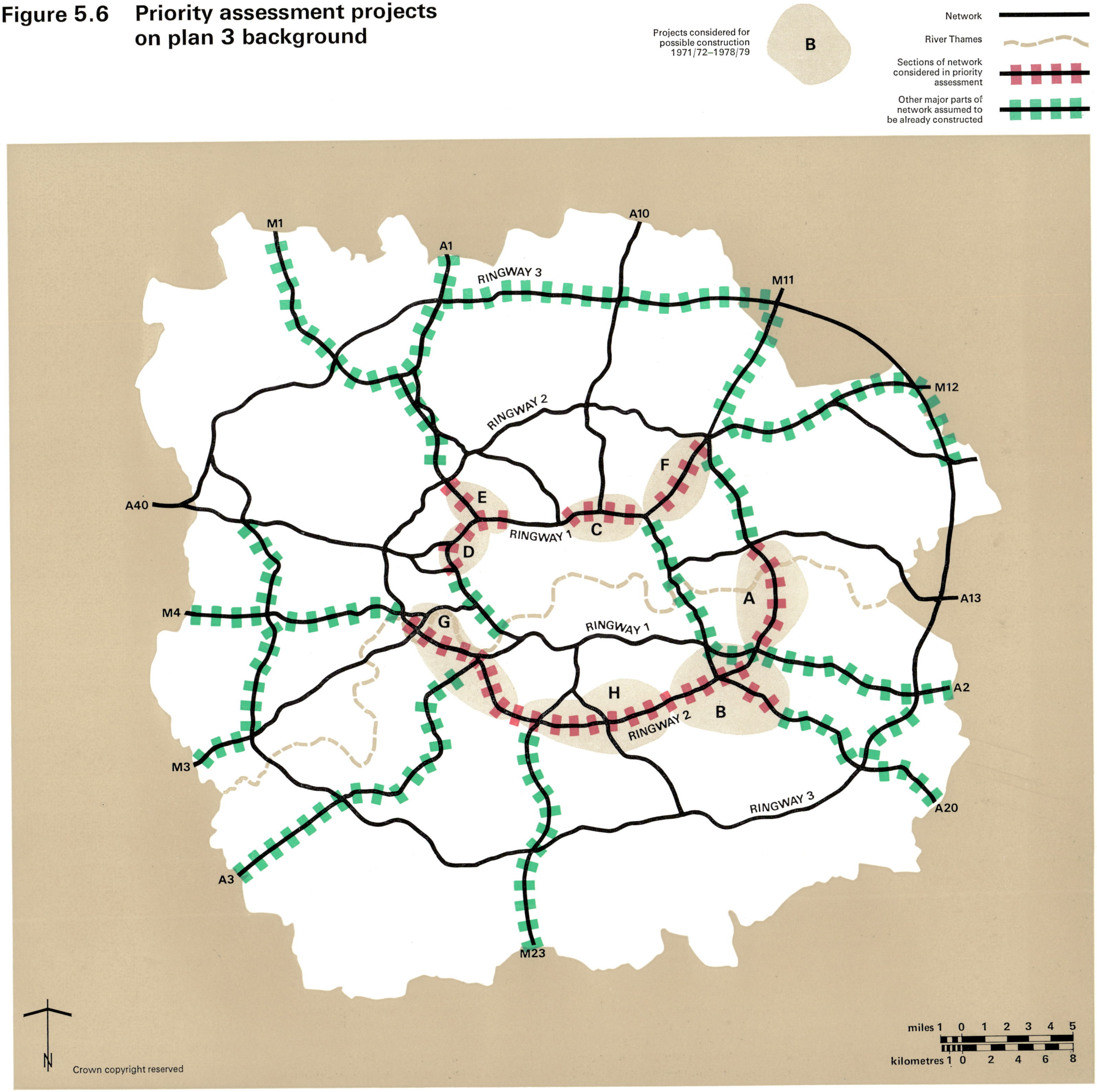

A

Appendix A Details of the Phase III model

General

A.1 The Phase III model was an extension and refinement of current transportation models. Particular emphasis was laid on incorporating public transport and assessing the effects of congestion and limitations in the capacity of the system on general levels of trip making. In this appendix extracts from or summaries of the consultant's description of the model are included. The emphasis is on differences from current methodology and there is no attempt to describe the whole model in detail. The first part deals with the model common to both public and private transport trips while the second part deals specifically with those aspects particular to the public transport network.

A.2 An overall flow diagram for the model is given in Figure 4.1 of the main report. It required two separate types of inputs, trip ends and networks. Techniques for calculating trip ends were used to determine generations and attractions by car-owners and non-car-owners separately. Distribution of car-owning trips (including some non-car-owning drivers) took place first, based on private transport travel times. Modal choice estimates were made using diversion curves which related the proportion of travel by public transport to the ratio of times by public and private transport. Speed change and trip restraint procedures were used to relate the total travel to the capacity of the system. The resultant road matrix was finally assigned to the network. The public transport trip ends were obtained by summing a proportion of the trips removed from the private matrix by the trip restraint procedure, trip ends from non-car-owning households and trip ends from car-owning households in the modal choice calculation. The total trip ends were distributed and then assigned to the public transport network.

Trip distribution (calibration)

A.3 It was assumed in the distribution process that the number of trips generated in one district and attracted to another district was a function of the total number of trips generated in the first district, the total number of trips attracted to the second district and the travel time between the districts.

A.4 These functions, known as distribution functions, were calculated from survey data and then assumed to remain the same for the design year. In this study six sets of functions were obtained for the three trip purposes (work, other home-based and non-home-based), for both car-owner and public transport trips.

A.5 To obtain the functions, the trip time range was split into a number of intervals such that roughly equal numbers of trips fell in each interval. It was found that thirteen intervals gave a convenient sample size. In the following formulae trip times refer to the midpoints of the intervals and are shown in Table A.1.

A.6 From the household interview data distribution functions F_{kt} for each district of attraction, k, and time interval, t, were calculated.

$$F_{kt} = \frac{\sum_{i(t)} T_{ik}}{A_k \sum_{i(t)} G_i}$$

i = district of generation

T_{ik} = trips generated in district i and attracted to district k

A_k = total trips attracted to district k

G_i = total trips generated in district i

$\sum_{i(t)}$ = summation over all districts of generation falling in time interval t

A.7 When these functions were plotted on a logarithmic scale against time, they approximated to straight lines, but with a discontinuity of gradient at a time of about twenty-five minutes. Thus the functions could be represented by the equations of the form:

$$F_{kt} = C_1 B_1^{-t} \quad \text{if } t < 25$$

$$F_{kt} = C_2 B_2^{-t} \quad \text{if } t > 25$$

A.8 These functions were needed for each district, but the amount of data available for individual districts was insufficient to enable all functions to be fitted. Thus as a preliminary step, they were obtained at sector level. This was done by plotting the data logarithmically and drawing the best lines by inspection.

Table A.1 Mid-points of time intervals to define distribution functions

Private transport (minutes)	*Public transport (minutes)*
5·0	20·0
8·0	28·0
9·0	31·0
10·0	34·0
11·5	37·0
13·5	40·0
15·5	43·0
18·5	46·0
22·5	49·0
28·5	53·0
38·5	58·0
54·5	66·0
75·0	80·0

A.9 In the small number of cases where an individual district had no data for more than five time intervals, its distribution function was assumed to be the same as that for its sector. The main use of the sector functions was to establish for each sector the relative gradients (H) of the two parts of the function, namely

$$H = \frac{B_1 - 1}{B_2 - 1}$$

This equation contains $B_1 - 1$ and $B_2 - 1$ rather than B_1 and B_2 because the gradient of the logarithmic distribution function is zero when $B = 1$. It was assumed that each district had the same value of H as the sector in which it fell. Additionally, the value of B for a district was assumed to differ from that of its sector by not more than 0·1.

A.10 In fitting the function to the district data, it was assumed that it passed through one of the points given by the data for the thirteen time intervals. It was also assumed that the value of B differed from that of the sector by a multiple of 0·01. This gave a total of 13×21 alternatives, assuming no zero samples (the points for which were ignored). The best fit was that which gave the smallest mean error for each time interval, the error being defined as:

$$\left| C B^{-t} - F_{kt} \right| \cdot \sqrt{\frac{A_k \sum_{i(t)} G_i}{C B^{-t}}}$$

Zero samples were not included in the error calculations.

A.11 In the calibration process, the base year trip ends were distributed using the functions found as described above. For each sector to sector movement the ratio L of the observed to the computed trips was found. Table A.2 lists some of the values obtained. The distribution functions were then multiplied by these factors and a further distribution made. At this stage the observed and computed trips were found to be sufficiently close in terms of the criteria used such as trip time distributions and screen line crossings. Agreement was regarded as satisfactory if average trip times were within 0·5 mins, screen line crossings within 10 per cent and somewhat larger margins on more severe tests such as particular parts of trip time distributions. The fact that the 'L' ratios were not all equal to unity indicates that, for a given place of attraction, the distribution functions vary somewhat according to the place of generation.

Table A.2 Ratio of observed to computed sector to sector movements (L) —all purposes

Mode	*Traffic sector*	0	1	2	3	4	5/6	7	8	9
Private	0	1·02								
	1	1·07	0·99							
	2	1·09	1·05	1·00						
	3	1·08	1·07	1·43	1·01					
	4	1·08	1·00	1·11	1·06	0·97				
	5/6	1·13	1·02	1·13	1·13	1·02	1·04			
	7	1·07	1·04	1·60	1·05	1·05	1·05	0·95		
	8	1·17	1·32	1·27	1·06	1·12	1·04	1·02	1·00	
	9	1·12	1·06	1·25	1·22	1·09	1·00	1·06	1·12	0·99
Public	0	0·98								
	1	1·06	0·98							
	2	1·04	1·06	0·96						
	3	1·03	1·08	0·99	0·99					
	4	1·04	1·04	1·07	1·00	0·98				
	5/6	1·02	1·03	1·01	1·01	0·99	0·98			
	7	1·08	1·06	1·05	1·03	1·04	0·99	0·98		
	8	1·13	1·16	1·33	0·98	1·10	1·01	1·06	0·97	
	9	1·13	1·18	1·33	1·09	1·07	1·00	1·09	1·20	0·96

Trip distribution (forecast)

A.12 The forecast of trip distribution was carried out separately for each of the three trip purposes, and for driver and public transport trips. In order to obtain the distribution of driver trips, the starting point consisted of all car-owner trip ends. A percentage of the generations in each district, namely the intra-district trips, was removed. These percentages were assumed to be the same as in the survey year.

A.13 The first iteration of the distribution was obtained by taking the number of trips T_{1ik} from district i to district k as

$$T_{1ik} = \frac{F_{ik} \cdot A_k \cdot G_i}{\sum_k (F_{ik} \cdot A_k)}$$

F_{ik} = distribution function for trips from district i to k using 1981 trip times

A_k = forecast attractions in district k

G_i = forecast generations in district i

A ratio R_1 was then found for each district of attraction

$$R_{1k} = \frac{A_k}{\sum_i T_{1ik}}$$

A second estimate of the trips was then made

$$T_{2ik} = \frac{F_{ik} \cdot R_{1k} \cdot A_k \cdot G_i}{\sum_k (F_{ik} \cdot A_k \cdot R_{1k})}$$

$$R_{2k} = \frac{R_{1k} \cdot A_k}{\sum_i T_{2ik}}$$

This was done a third time, the criterion for convergence being that the average (unsigned) error should be within the range 2 to 5 per cent.

Modal choice

A.14 The matrix obtained was then split between driver and public transport trips using diversion curves which gave, for each ratio of trip time by public transport to trip time by private transport, the proportion of trips using public transport. For intra-district trips, the modal choice was assumed to be the same as it was for each district in the survey year.

A.15 Trip attractions of driver mode by non-car-owners were taken and added into the matrix in the same proportions as the trips already made from each district. The remaining non-car-owner trip ends from the generation procedure were added to the car-owner public transport trip ends obtained from the modal choice procedure. These were then distributed using a distribution model calibrated in the same way as for private transport, but using the distribution functions derived from data on public transport trips.

A.16 The diversion curves were obtained from the household interview data for each of the three trip purposes in the following way. A public transport skim tree (i.e., a matrix of district to district journey times) was built for 1963. The times were taken as the lesser of rail and bus times. The former was derived by using a BPR network, while the bus times were 10 minutes plus 1·8 × (off-peak road time).

A.17 The percentage of car-owner attractions in each sector travelling by public transport was calculated for each of the time ratios (public trip time/private trip time) 0·6 to 2·5, at 0·1 intervals. The trips were divided into long and short trips, the dividing line being a trip time of 15 minutes by private transport, excluding parking. Three minutes was added to each private trip time to allow for parking and an extra 3 minutes in the Central Area. The two Central Area sectors were combined to form one.

A.18 The 54 curves (3 purposes × long, short × 9 sectors) thus obtained were plotted and smoothed curves fitted by observation. For the forecast of modal choice, where the samples were large the actual figures obtained were used, and where they were small the values of the smoothed curves were used. Three representative curves are shown in Fig. A.1.

A.19 The output, after application of the diversion curves to the matrix of all car-owner trips for 1981, consisted of both private and public transport trips. The private transport trips are now discussed while the treatment of public transport trips is discussed later.

Speed change

A.20 It was found that when the road trips were assigned to the future networks the volume of traffic on parts of the system exceeded the capacity.

A.21 A combination of two methods of reducing traffic to network capacity was used, by reducing motorway speeds (speed change) and by removal of trips from the system using linear programming (trip restraint). The manner in which the linear programming was applied removed approximately the same amount of traffic from the motorways as from the base roads. Thus, if both types of road were to carry the maximum possible amount of traffic after the linear programming, then prior to this process they had to have overloads which were approximately equal.

A.22 This was achieved by the process of reducing motorway speeds, which altered the relative loading of the motorways and the base roads. This balancing of overloads was done separately for each of four large areas (jurisdictions) which made up the LTS area. After each speed change, the distribution and modal choice procedures were repeated as well as the assignment. Thus the speed reductions caused reduction in volumes on overloaded links by a combination of three methods, redistribution, transfer to public transport and possible re-assignment of road trips to alternative routes.

A.23 In balancing overloads, the procedure equated the overload on overloaded motorway links with the total overload on all the base roads in the surrounding jurisdiction. Thus it was assumed that, whereas some motorway links might be less than full even in conditions of excess demand, the base roads would be completely full in the sense of their traffic loading equalling defined capacity.

A.24 The procedure described below calculated a target volume for motorway links in each jurisdiction, which was the level that equated their overloads with those of surrounding roads, to

Figure A.1
Specimen diversion curves

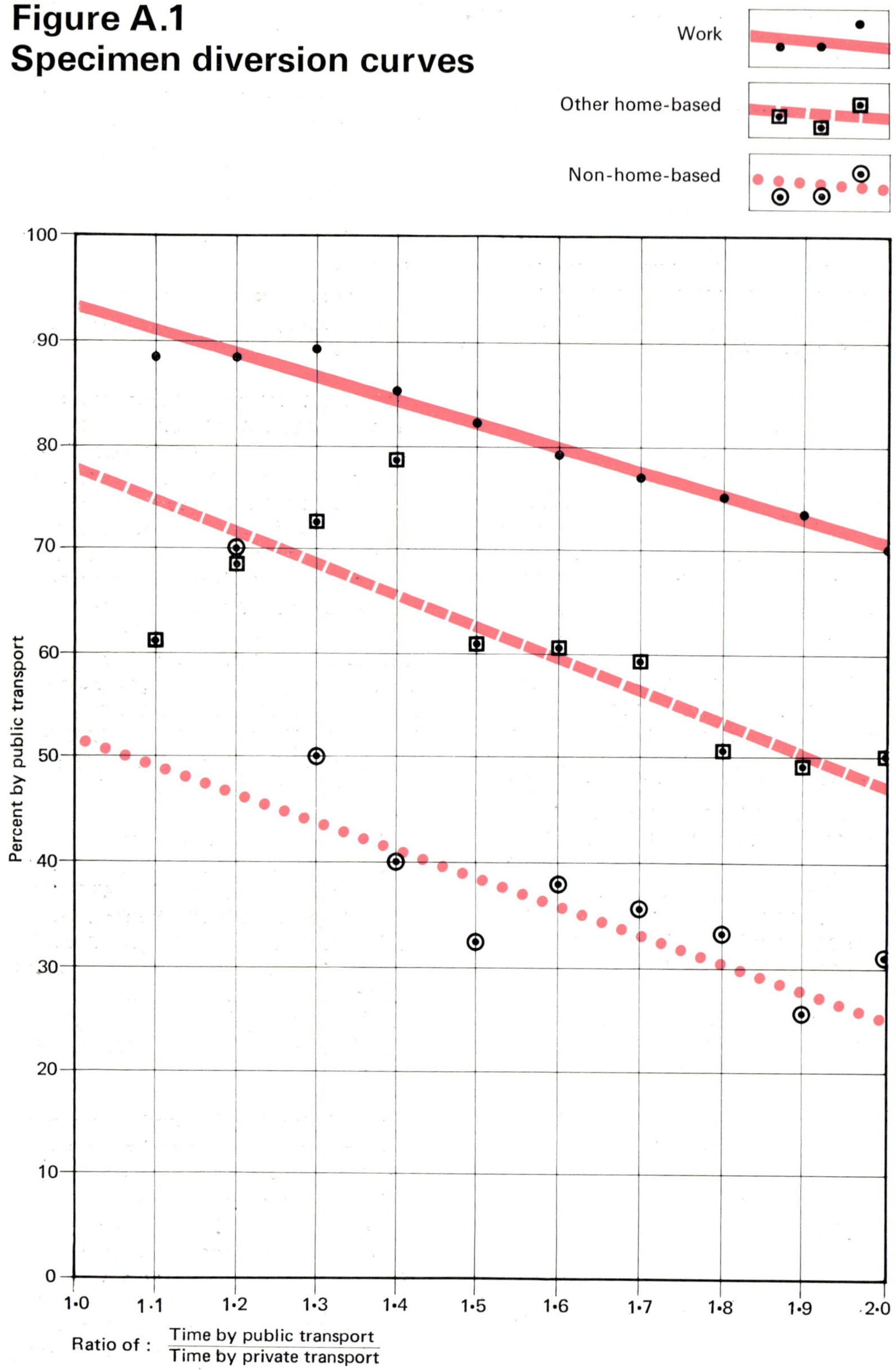

give the assignment prior to the linear program. Let

M = the sum of loadings before speed change on overloaded motorway links

T = target motorway loading (thousands of vehicles per day)

N = number of overloaded links

k = a constant

G_1 = vehicle miles on base roads before speed changes

G_2 = vehicle miles on base roads after speed changes

C = defined capacity of base roads within a jurisdiction (vehicle miles per day)

then it was found empirically that:

$$M-TN=k(G_2-G_1)$$

A.25 This reflected the fact that loadings on the base roads increased as those on the motorways decreased. As the linear program removed about equal proportions of traffic from motorways and ordinary roads and as the aim was to have both the base roads and the previously overloaded motorways running at capacity after the linear programming, it was required that:

$$G_2/C=T/150$$

The figure 150 corresponds to the assumption that the 'limiting volume' of a dual four-lane motorway is 150,000 vehicles per day. Eliminating G_2 between the two equations given above:

$$T=\frac{M+kG_1}{kC/150+N}$$

This gave the target volume, from which the link speeds could be found.

A.26 The method of finding the speeds that would give these volumes in the assignment process is described below, with the suffixes 1, 2 and 3 being used for the three distribution and assignment stages that preceded the linear program.

A.27 For each jurisdiction, the first estimate of the target volume was made, using the formula shown in paragraph A.25:

$$T_1 = \frac{M_1 + k_1 G_1}{k_1 C/150 + N_1}$$

Calculations were then made of the second speed for each of the links having volumes greater than T_1. Then S_2 was given by the empirical formula:

$$S_2 = S_1 - 30\,(1 - \sqrt{T_1/Q_1})$$

where S_1 was the first speed on the link and Q_1 the first volume. Traffic was then redistributed and reassigned to give new volumes. k_2 was calculated (all jurisdictions combined) from:

$$k_2 = (M_1 - M_2)\,/\,(G_2 - G_1)$$

New target volumes were calculated. There were then N_2 overloaded links:

$$T_2 = \frac{M_2 + k_2 G_2}{k_2 C/150 + N_2}$$

The speeds S_3 were then calculated by interpolation between the first two speeds for links previously overloaded:

$$S_3 = \frac{(T_2 - Q_2)\,(S_1 - S_2)}{(Q_1 - Q_2)} + S_2$$

If there were any links on which the volume had risen, the previous empirical formula was used.

A.28 The next assignment was now made. It was judged that the flows thus obtained were sufficiently close to the target volumes for further iterations of the speed change procedure to be unnecessary.

Trip restraint by linear programming

A.29 The purpose of the linear programming routine was to maximise the number of internal private trips able to take place subject to no motorway links being left overloaded. The linear program removed, for each district of attraction, a certain proportion of the trips, this proportion not depending on the district of generation. This feature of the model removed some potential traffic that did not, in fact, want to use any of the overloaded links.

A.30 For each road plan a maximum was imposed on the proportion of trip attractions of each district which could be removed by the linear program. One effect of this was slightly to reduce the overall number of trips able to use the system. However, it was thought unrealistic to allow too large a difference between the proportions of trips restrained for different district groups.

A.31 The LTS area was divided into 33 groups of districts and a set of overloaded motorway links chosen so that, when the traffic on these links was reduced to their capacity, all the other links in the network were also working within their capacity. In effect these links were the bottlenecks. Commercial vehicle, taxi and external trips were regarded as being essential traffic and were not subject to the restraint procedure.

A.32 In its simplest form the linear programming procedure can be expressed as follows:

$$\text{Maximise} \sum_{x=1}^{n} A_x$$

subject to

$$\sum_{x=1}^{n} k_{rx} A_x \leqslant T_r \quad r=1,\ldots m \quad (1)$$

$$A_x \leqslant B_x \quad x=1,\ldots n \quad (2)$$

$$A_x \geqslant pB_x \quad x=1,\ldots n \quad (3)$$

where

A_x = allowed internal private attractions to district group x

B_x = desired internal private attractions to group x

k_{rx} = the fraction of trips attracted to group x which pass through link r

T_r = capacity on link r available for private trips, i.e., (total capacity of link)—(capacity used by commercial vehicle, taxi and external trips)

m = number of selected links

n = number of groups of districts

(1-p) = maximum allowed restraint in a group.

A.33 Constraints (1) express the condition that the capacity used on link r by internal private trips must not exceed the capacity of the link less the capacity used by commercial, taxi and external trips. The coefficients k_{rx} are assumed to be independent of the values of A_x. Constraints (2) express the condition that the allowed attractions to each group must not exceed the desired attractions, that is there must be no more trips attracted to any group after the linear program than before it. Constraints (3) prevent the number of allowed attractions to any group from being less than the imposed limit.

A.34 The solution to this set of equations gives the private trip attractions allowed in each group of districts which, together with the commercial vehicle, taxi and external trips, do not overload the network. There were in fact small errors in the target volumes calculated during the speed change procedure which did leave some small overloads.

A.35 The linear programming method described was regarded as the standard one and was applied in this form to road plan 3. In the case of road plan 9 no linear programming was required. After the necessary speed change had been made it was found that the system was working within its defined capacity. In the case of road plan 1, there was insufficient motorway for the standard method to be used. Thus the method was modified, so that the vehicle mileage on base roads in each jurisdiction was restricted to a calculated capacity. As before the capacity used by commercial vehicle, taxi and external trips was subtracted from the calculated capacity to give a capacity available for internal private trips. However commercial vehicles, taxis and external trips were reduced as a separate operation before the linear programming routine because the lack of motorways compared with plan 3 would cause the total vehicle usage to be lower.

A.36 Of the trips removed from the road system by the linear programming, some were transferred to public transport and some suppressed. The proportion transferred was based on the assumption that car-owners prevented from using their cars would behave like non-car-owners. Hence the proportion transferred was the ratio of the differences in public transport and driver trip rates, for car-owning and non-car-owning households, that is:

$$\frac{\text{NCO public transport—CO public transport}}{\text{CO driver—NCO driver}}$$

From 1962 survey data, these proportions were found to be:

Work 0·735

OHB 0·273

NHB 0·153

Thus most of the work trips go to public transport, but only a minority of non-work trips.

Public transport

General

A.37 The objective of the techniques developed for public transport analysis was to devise realistic matrices of trips by public transport (both in 1962 and in 1981) and to develop a process of assigning them to alternative networks.

A.38 The most important technique relating to public transport analysis developed in Phase III was a new method of route description and shortest path determination which was termed Transitnet. Use was made of Transitnet in most stages of the public transport model so it is described first, and descriptions of matrix compilation, trip assignment and network design follow.

The route finding algorithm

A.39 Public transport networks differ fundamentally from road networks by comprising a system of fixed services following given routes, generally at fixed intervals. In Transitnet, links and nodes were defined in the conventional manner but the network was built up by the description of individual public transport services, each service being described by the nodes it followed, the time or speed between nodes, and the frequency of operation. A special algorithm was used to determine the shortest path between any pair of zone centroids in the network described. The criterion for deciding on the path was that of minimum time. The journey time included both link travelling times plus waiting times at interchanges and walking times. If required, additional notional time penalties could be added every time an interchange took place in order to simulate preference for taking a slightly longer path to avoid a change. These were not in fact used in Phase III. Having calculated the shortest paths through the network in terms of time, these times were used for the distribution of trips between zone pairs and then to establish the paths for the assignment of trips to the network, it being assumed that all trips would be made by the shortest path available.

A.40 The calculation of paths through a network was carried out by the computer in a number of simple steps, the complexity of the overall network being reflected in the total number of steps involved. The computer first selected one node as a 'home' node and calculated the paths from the home to all other nodes in the network. This was known as the tree for that particular home node. Having built the tree for the home node selected, the process was repeated using each zone centroid in turn as a home node.

Public transport trip matrices

A.41 Public transport trips were obtained during the model procedure from:

a. The majority of the trips made from non-car-owning households.

b. The modal choice procedure.

c. The trip restraint procedure.

The public transport internal trips resulting from the above procedures were divided into three purposes, work, other home-based, non-home-based.

A.42 When the London Traffic Survey was started in 1962, it was not envisaged that an analysis of public transport would be included in the project and consequently the terms of reference did not allow for the collection of satisfactory public transport trip data or for checks of the amount of traffic actually carried on the public transport system in order to calibrate and check the model. When it was decided to enlarge the scope of the study to include a more detailed analysis of public transport it was not practicable to collect much additional information on the subject. Available information indicated that the disparity between estimates of the number of trips made in London and those recorded in the home interview or postcard survey was not small enough to be ignored. It was decided to estimate an 'unreported trip matrix' based upon the information available.

A.43 The readily available information consisted of the internal public transport trips made by residents of households, obtained from the home interview study, and a matrix of trips made by medium-distance external rail passengers between the arrival-at-destination times of 8 am and 10 am. This was based on the postcard cordon survey, supplemented by equivalent 24-hour figures from selected control stations. In addition, the British Railways and London Transport Boards were able to supply figures of the number of passengers carried on their services, although these sometimes related to years other than 1962.

A.44 The process of synthesising unreported trip matrices for 1962 from this information consisted of three stages:

a. Estimation of control totals for trips by public transport. This involved adjusting the Boards' estimates of trips to make them compatible with the LTS estimates of total journeys by individual modes.

b. Estimation of numbers of trips by mode and by class of trip (such as long distance rail trips by non-residents). This involved estimating and listing total numbers and modal choice (bus, Underground, British Rail) of classes of journeys both reported and those known to have been unreported in the LTS. They were then compared with the figures in **a** and adjusted to bring the total trips by mode as closely into agreement as possible.

c. Derivation of matrices. This involved devising matrices suitable for each class of trip enumerated in **b** and was done mainly by comparison with the distribution of similar classes of reported trips, that is by selectively factoring appropriate matrices of reported trips.

The public transport matrices used for assignment were then obtained by the addition of the reported and unreported trips.

Peak hour estimates

A.45 Owing mainly to the influence of work trips, the proportion of daily trips carried during peak periods by public transport is considerably higher than the proportion by private transport and it is peak hour demand which determines capacity requirements. This resulted in the decision to estimate peak hour travel by public transport in addition to 24-hour travel. It was decided that peak hour matrices should be produced by factoring the 1981 24-hour matrices, which were in the form of generations to attractions, to provide peak hour matrices in origin to destination form. The factors were based on peak period trips, determined from the 1962 household interviews, expressed as a proportion of 24-hour trips. The morning peak period was chosen primarily because the 10 minute peak within a peak period, on which capacities were based, is higher in the morning than in the evening. It was found that the peak arrivals occurred in the Central Area between 8.40 am and 9.40 am, but in the non-Central Area between 8.20 am and 9.20 am. It was decided to use these separate periods according to destinations. It will be noted that the 8.40 am to 9.40 am Central Area arrivals pass through the suburbs about 20 minutes earlier. The time displacement of the Central Area trips gives a better estimate of the suburban loadings as well as the Central Area loadings than would the use of the 8.20 am to 9.20 am timing everywhere.

A.46 Consideration was first given to using zone-to-zone peak hour factors, but it was found that so many zone pairs in 1962 had either zero peak hour trips or unreliably small sample sizes that some grouping of zones was necessary. It was decided to use sector-to-sector factors, but even with areas of this size some samples were found to be too small and, for example, cross London trips had to be grouped and averaged to provide reliable factors for this kind of trip. It was however possible to subdivide Sector 5 into 'City' and 'West End' and deal with these sub-sectors separately. This proved well worthwhile because the nature of the two sub-sectors leads to significant variations in the peak hour factors, especially for other home-based trips.

Assignment to the network

A.47 The matrices of person trips, determined as described above, were assigned to networks using Transitnet. This provided total numbers of people travelling on each link under the given conditions. In some cases two assignments were made to the same network, one of peak hour trips and one of 24-hour trips. Peak hour assignments gave volume of traffic by direction on each link but the 24-hour assignments were based on generation-to-attraction matrices and were therefore non-directional.

B

Appendix B Details of alternative transport networks

Road networks

General

B.1 In formulating alternative road networks, two broad groups of plans were evolved. The first series of networks was based generally on the 1981A network defined for Phase II of the LTS and the related traffic estimates. The second series of networks was based on a more general analysis of travel demands and the formulation of a possible network to accommodate these demands. The following descriptions of both road and public transport networks are again extracts from or summaries of the consultant's report.

The first series of networks

B.2 The 1981A network was examined in some detail in Phase II and, as a result of the assignments made to it together with other planning criteria, a total of twenty-three alternative networks were formulated for consideration in Phase III. These networks varied in the extent to which movements in different parts of the LTS area were provided for, in the extent to which private or public transport was encouraged and in their land use implications. Some of these networks included a number of motorway routes within Ringway 1.

B.3 Seven of the twenty-three networks which had been formulated at this stage were chosen for further consideration. The seven networks chosen (plans 1, 2A, 3, 3H, 5A, 6 and 8) are outlined below, their motorway systems are shown in outline in Figure B.1 and their motorway mileages and possible orders of cost are given in Table B.1.

Table B.1 Comparison of motorway mileages and costs

Series	*Network*	*Miles on or within Ringway 3*	*Miles within Ringway 1†*	*Order of costs (Plan 3=100)*
	1981A	330	8	110+
First series	*Plan 1	142	0	22
	Plan 2A	152	0	50+
	*Plan 3	337	0	100
	Plan 3H	350	13	120
	Plan 5A	361	0	120
	Plan 6	367	0	100
	Plan 8	300	N/A	100
Second series	*Plan 9	407	0	125
	Plan 10	520	60	250
	10A	325	0	100+
	10B	322	13	110+
	10C	338	13	120−
	10D	360	27	130
	10E	420	27	140+

*Plans examined in district test series

†The Western Avenue extension to Marylebone Road and similar roads within Ringway 1 have not been included

Figure B.1 Road plans 1–9

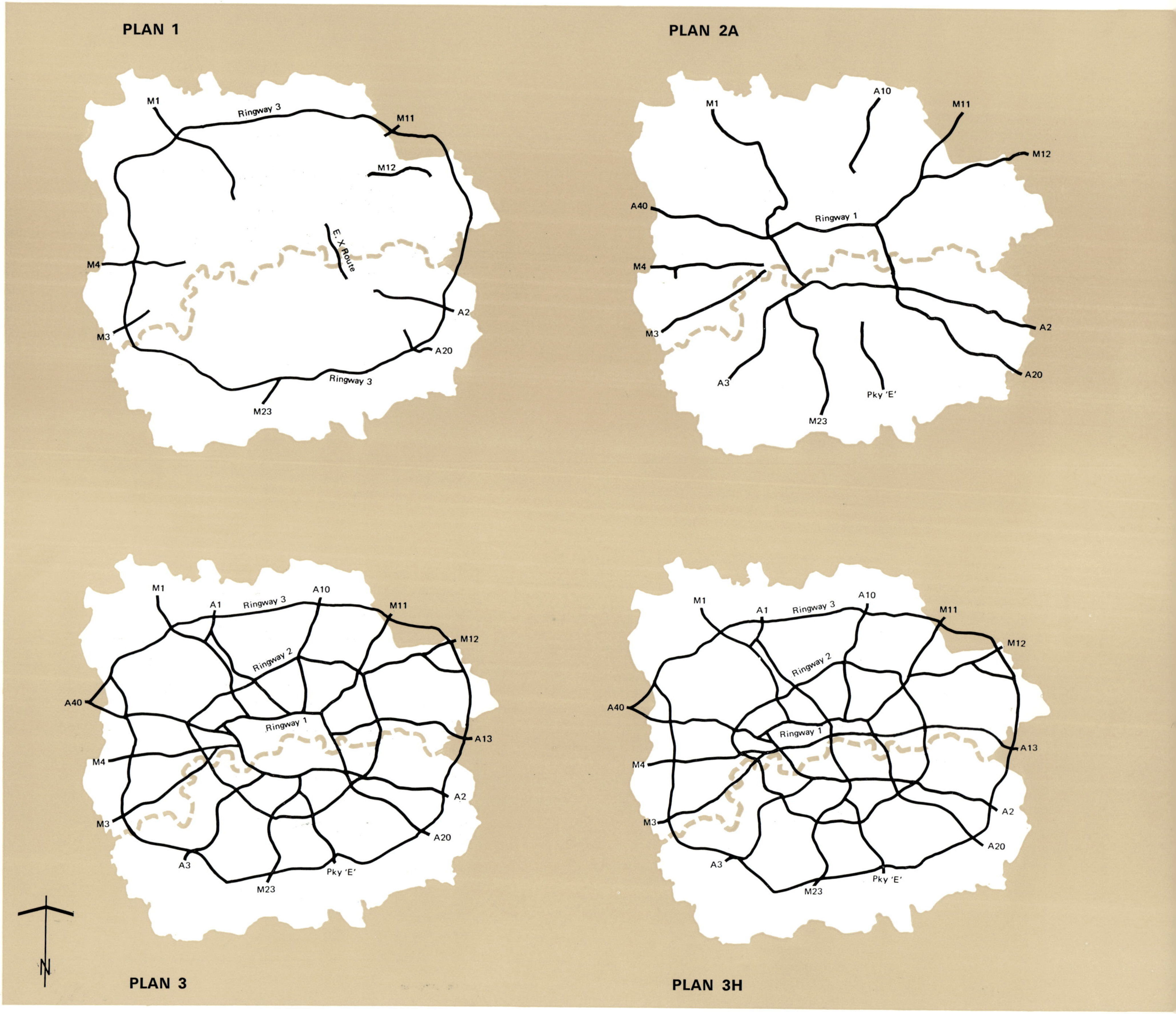

PLAN 5A

PLAN 6

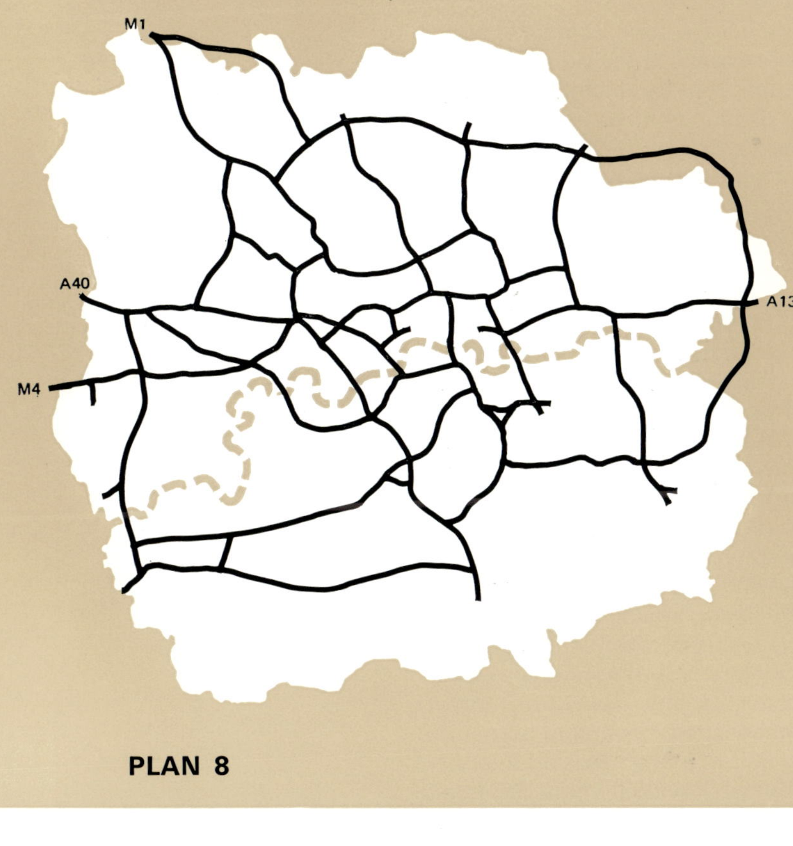

PLAN 8

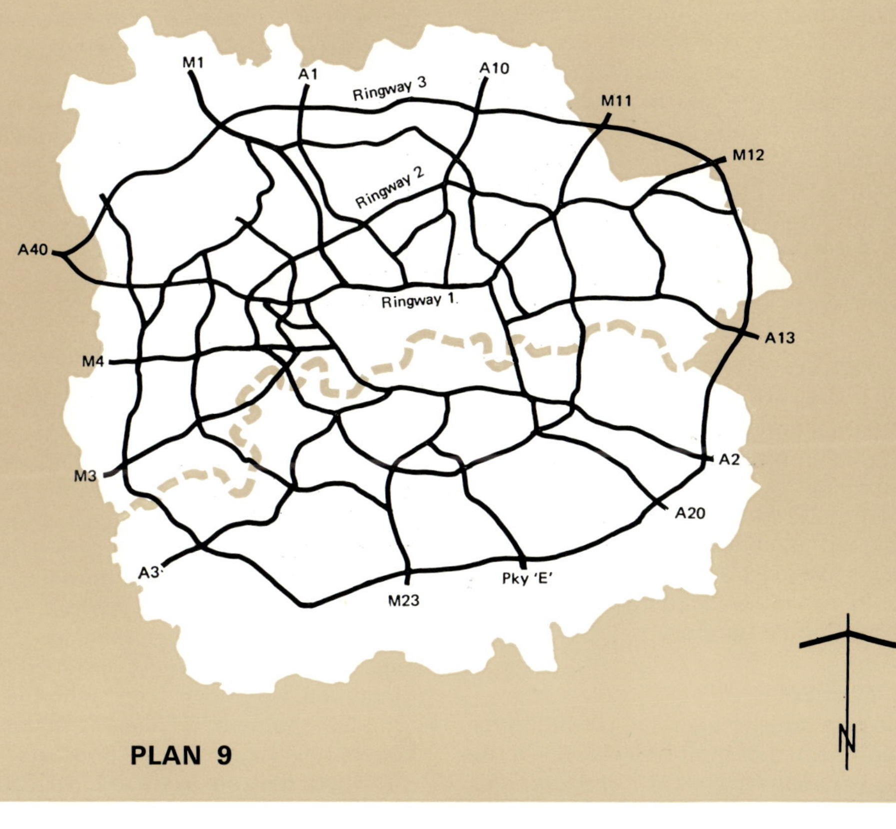

PLAN 9

B.4 These networks and plan 9, which is also included in Figure B.1 and Table B.1, are briefly described as follows:

a. *Plan 1*
This plan reflects a policy of limited expenditure on motorways with a resulting considerable degree of restraint in the use of private cars. It consists of Ringway 3 around London linking the national motorways and, in addition, those motorways likely to be constructed by the early 1970's as a result of the then current programmes of work at the time the plan was drawn up. It does not have a coherent system of motorways within Ringway 3, the fragmented motorways being linked through the present system of roads.

b. *Plan 2A*
This plan is oriented to serve cross-London traffic to a greater extent than plan 1. The main radials are brought in to the line of Ringway 2 but this is not constructed to motorway standard. Selected motorways are carried across London forming Ringway 1.

c. *Plan 3*
Plan 3 is formed by a comprehensive system of motorways linking with the national motorways. It is similar to the 1981A network and consists of three ringways and several radial motorways. It differs from the 1981A network primarily in that the Stamford Hill–St John's link is omitted, the M11 is re-aligned on the Roding Valley route rather than up the Lea Valley and Ringway 3 is re-aligned in the north-west and south-west and is completed in the south-west.

d. *Plan 3H*
This is similar to plan 3 but contains two additional motorway routes crossing the area within Ringway 1, one in a north–south and the other in an east–west direction. These routes would be designed primarily for motorway bus access to and across Central London, but would also be available for car trips across Central London.

e. *Plan 5A*
This plan adds to plan 3 in providing for a high degree of motorway service in the area between Ringways 1 and 2 where a very large increase in motor vehicle usage is expected. The provision of additional routes in this area should encourage its redevelopment.

f. *Plan 6*
The redevelopment of areas between the Ringway 2 and the line of Ringway 3 is encouraged in this plan. The configuration of motorways is similar to that in plan 3, but Ringway 3 is excluded and the lines of certain radials are modified.

g. *Plan 8*
This provides an irregular pattern of motorways chosen to follow lines of special opportunity such as comprehensive redevelopment areas, vacant sites and abandoned railways. The pattern formed tends to a grid rather than ring-and-radial system.

h. *Plan 9*
Plan 9, although based on the first series of networks, was formulated making use of the information obtained from the derivation of plan 10 and the assignment to that plan. Taking plan 3 as a basis, an attempt was made to add motorways so as to relieve the most severe cases of overloading found from assignments to plan 3. This overloading had been found to be chiefly in the north and west. The analysis undertaken in deriving plan 10 confirmed the large travel desires in these areas. Accordingly, motorways are added to those in plan 3, primarily in the north and west. The additional motorways are made as continuous as possible, the greater part of those added forming part of an additional ring between the lines of Ringways 2 and 3.

The second series of networks

B.5 The networks described above were in many cases based on the general conclusions drawn in Phase II about the probable increases in trip generation and on the results obtained from assigning traffic to the 1981A network. However, in formulating networks for London and examining the traffic assigned to them, the designer is faced with the problem that within limits, no matter where new high-class facilities are provided, they are likely to have a considerable amount of traffic assigned to them by reason of their superiority over the existing system of roads. Although comparison of measures such as total vehicle-hours and vehicle-miles gives some indication of the relative overall efficiency of alternative systems, it does not indicate whether parts of a system are located so as to serve best the desires of traffic. Plan 10 and its variants was drawn up in an attempt to take these traffic desires into account more directly. The method adopted is described below.

B.6 The movements between 50 traffic areas formed by combining zones were examined using the trips forecast for 1981 and the distribution of these trips obtained with the 1981A network. The straight-line traffic desires between area centroids were then summed across sections of an arbitrary square grid of screen lines. The resulting movements across these screen lines indicated the total desire for movement. In order to obtain a measure of the proportion of movements that might reasonably use motorways with provision for substantial movement by private car, reference was made to American data relating to networks planned for large cities in that country. Such data indicated that some 45 per cent of vehicular trips made in a large city might be expected to use motorways. In the present study it was found that by eliminating all trips with a distance of less than four miles between area centroids, the remaining trips constituted 45 per cent of total trips. The sum of these remaining trips across each section of the grid was then translated into motorway lanes.

B.7 The data gave little indication however of how such motorways could be best connected to form a system. It was thus necessary to use screen line information as a guide to defining a system. Accordingly checks were made that sufficient total capacity had been provided across screen lines formed by natural barriers, such as rivers and railway lines. The approximate capacity of the existing streets crossing these barriers was estimated and added to the capacity of any motorways provided. This total capacity was compared with the total desired movement across the barrier and in a few cases additional motorways provided where necessary.

B.8 Plan 10, the motorway network drawn up using the approach described above, is shown in outline in Figure B.2. It consists of some 520 miles of motorway route on or within Ringway 3 (compared with about 340 miles in plan 3) and a rough estimate of its cost was some two-and-a-half times that of plan 3 as shown in Table B.1. It contains much of the plan 3 rings but differs from plan 3 in the orientation of many routes and contains a considerable mileage of additional motorway routes, particularly in the centre and in the west and north-west of the LTS area.

B.9 Plan 10 provides a useful indication of the extent of network which might be necessary to accommodate the 1981A travel demands to a similar degree on motorways to that envisaged in large American cities. It also indicates which parts of the LTS area require a high density of motorways to satisfy demand in full and those parts where the forecasts have revealed little demand for motorway-type facilities. However, the extent and cost of the full plan 10 motorway system was extremely large and it was thus necessary to produce some motorway networks based on plan 10, but more comparable in scale with some of the first series of networks. In order to do this and to obtain a rough check of the validity of the method adopted in drawing up plan 10, the complete plan 10 network was coded (at a district level) and an assignment of 1981 travel made using this network. The results of the preliminary assignment were approximately those expected when developing plan 10 and were used as a guide in formulating the reduced versions of the complete network.

B.10 Five reduced versions of plan 10 were produced; their approximate motorway mileages and costs are shown in Table B.1 and their approximate motorway locations in Figure B.2.

Networks tested in Phase III

B.11 Limitations of budget and time restricted to three the number of alternative road networks which could be tested. In view of the importance of determining the effect of very different levels of investment in a road system, as much as possible was kept constant in each road plan tested, the main difference between plans being the extent of the motorway system in each. The three plans selected, plans 1, 3 and 9 were all based on the first series of networks and all had common motorway routes. Plan 3 had routes additional to those in plan 1 and plan 9 likewise had additional motorways to those in plan 3. The costs and mileages of these networks are shown in Table B.2.

Table B.2 Costs* and mileages of motorways in tested networks

Road network	*Cost*	*Miles*
Plan 1	372 (22%)	150 (43%)
Plan 3	1658 (100%)	347 (100%)
Plan 9	2063 (125%)	413 (119%)

*Costs in £ millions (at 1966 prices)
[Plan 3=100%]

Public transport networks

General

B.12 In Phase III changes in the public transport system were aimed largely at improving the level of service provided and this was to be effected by increases in the capacity of overloaded links and a reduction of transport interchanges. A number of possible plans was considered before the networks for testing were chosen. British Rail services are considered first followed by Underground and bus services. In both cases the plans outlined do not constitute final proposals. They are hypothetical and do not necessarily represent current strategies but, as with the road plans, they form a basis for examination, testing and discussion.

Figure B.2 Road plans 10 and 10A–E

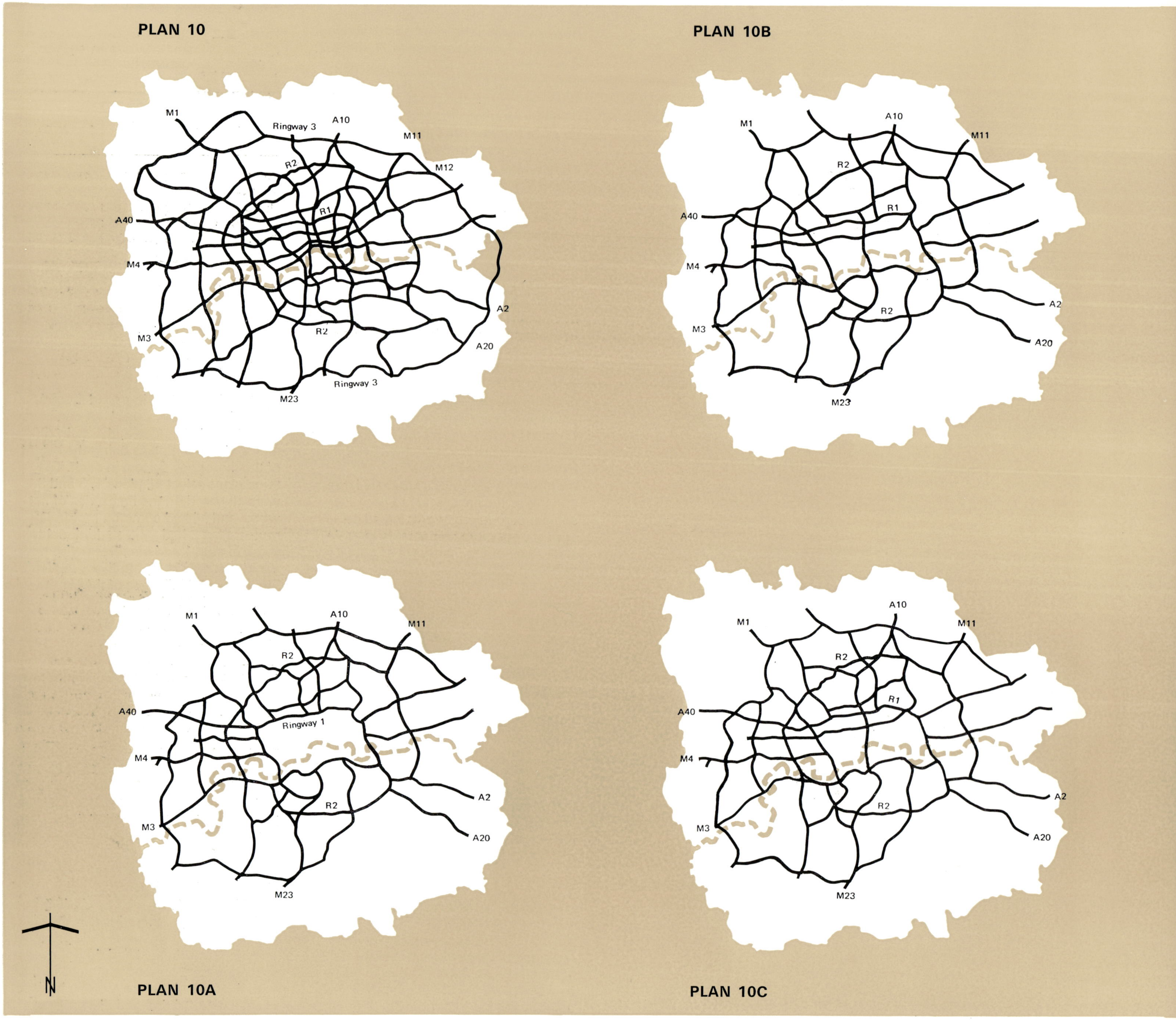

PLAN 10D

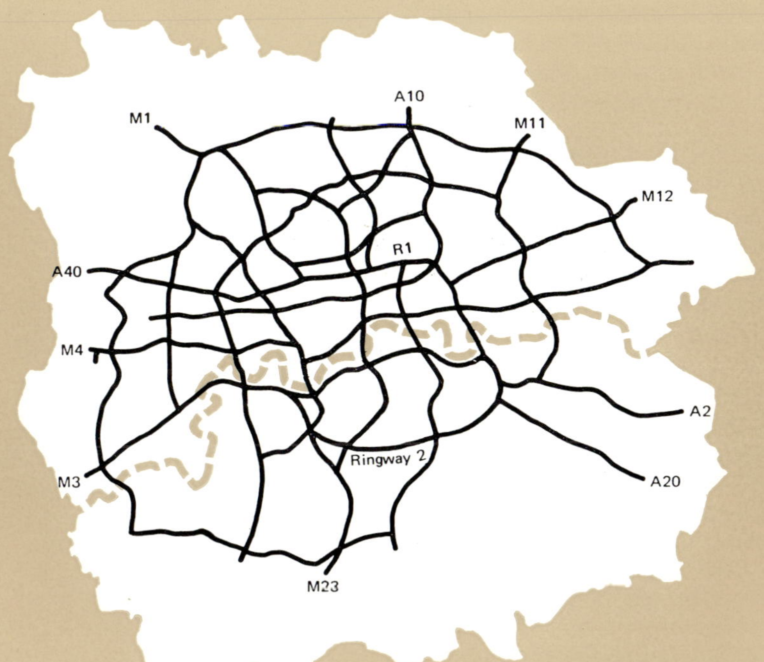

PLAN 10E

British Rail services

B.13 British Rail plans A1, B1 and C1, based upon the existing system, form a series with an increasing level of service. They all envisage the electrification of the Great Northern line to Hitchin and Royston, together with the electrification of the Hertford loop to Hertford North and Stevenage. Suburban services starting from Hertford North and Welwyn Garden City are diverted at Finsbury Park through the Northern City line to Moorgate, while services from Hitchin and Royston continue to operate into Kings Cross. The plans also provide for the electrification of the Lea Valley line from Clapton to Cheshunt, the electrification of the existing line linking Blackfriars with Farringdon and of the Metropolitan widened lines between Farringdon and Kings Cross/St Pancras to enable Southern Region trains to run through to the northern suburbs of London, and the construction of a new branch from Feltham to London Airport (Heathrow) to provide an express service to Victoria. The Euston–Watford–Crewe line was electrified after 1962 and was included in all rail networks. Plan B1 in addition includes an increase in track capacity at Borough Market Junction, East Croydon and New Cross Gate. Plan C1 also provides for the electrification of the Midland line from Bedford to St Pancras; the connecting link between Kentish Town and Farringdon; the transfer of the Leyton to Ongar and Woodford to Hainault lines from London Transport to British Rail; the diversion of some Shenfield services from Liverpool Street to Fenchurch Street and the introduction of a service from Clapham Junction via Olympia to Paddington. Additional track capacity is provided between Shortlands and Peckham Rye, and at Battersea. Certain infrequently used branches are closed down in each of the plans, more in A1 and less in C1. British Rail plan C2 is the same as C1 except that the Ongar and Hainault services are not transferred to British Rail, so as to be compatible with Underground plan B. This enables all the Shenfield services to be left at Liverpool Street instead of some being diverted to Fenchurch Street. Plans C1 and C2 also envisage a general improvement to the Southern Region services. For instance, 1962 journey times are reduced by up to 40 per cent and by an average of about

15 per cent. There is also an increase in the number of services into British Rail termini. These service improvements are achieved by the track improvements mentioned above and by increased motive power.

B.14 Plan D, which was prepared by the consultants independently of British Rail, was the first to employ through main line links. The idea was to form a railway 'box' with two additional cross links. The passenger would continue on a main line train to a station where he could interchange to a suitable distribution service which would take him to his final destination. The northern east–west link coincides with the Metropolitan Line between Paddington and Liverpool Street and the southern links Waterloo through to London Bridge involving the closure of Cannon Street, the withdrawal of main line services to Charing Cross and the replacement of the latter by a shuttle service from Waterloo. The north–south links are between Moorgate and North Kent East Junction via London Bridge; Farringdon to Blackfriars; Marylebone to Victoria via Marble Arch and Hyde Park Corner; Willesden Junction to Clapham Junction.

B.15 Plan E, also prepared by the consultants, again provides main-line through links. Some trains at present serving Waterloo are diverted via Nine Elms, across the river, and then in tunnel to St James's Park, Trafalgar Square, Aldwych, Ludgate, Cannon Street, Fenchurch Street and Surrey Canal. Services from Aylesbury and High Wycombe are diverted to Paddington and taken, with other Paddington local services, in tunnel via Oxford Circus, Leicester Square, Charing Cross, Waterloo and Bricklayers Arms to link with the Southern Region. Marylebone Station is closed. As the third cross link the Watford and Richmond services are diverted from Broad Street through St Pancras and Farringdon to Blackfriars and the South.

B.16 British Rail plan G consists of plan C2 with the addition of two through east–west links meeting each other with cross platform interchange at Covent Garden, and connecting with the north–south link (proposed in plan C2) from Farringdon to Ludgate. The northern east–west link diverts some Liverpool Street trains underground via a station between Liverpool Street and Moorgate, Ludgate (interchange with the north–south link), Covent Garden and Oxford Street to Paddington. The southern link runs from New Cross Gate via London Bridge, Cannon Street, Ludgate (interchange with the north–south link), Covent Garden, Piccadilly and Victoria to Clapham Junction. Plan G also includes the electrification of the main lines from Paddington to Reading and High Wycombe, and the Oxted branch of the Southern Region (Central Division). The Fleet Line branch to New Cross Gate is withdrawn.

London Transport services

B.17 A range of improved networks was considered in Underground plans A to D. Plan A is a minimum plan which consists of the existing network together with the Victoria line which is now complete between Walthamstow and Victoria. The Northern City line is handed over to British Rail.

B.18 In Underground plan B the Victoria line is extended to Brixton. The Piccadilly line is extended from Hounslow West to Heathrow and the Holborn–Aldwych branch to Waterloo. The Fleet line incorporates the Stanmore branch of the Bakerloo line and continues it from Baker Street via Bond Street, Green Park, Trafalgar Square, Ludgate Circus, Fenchurch Street, Surrey Docks and New Cross to Lewisham, with a branch to New Cross Gate.

B.19 Underground plan C comprises plan B plus the Bakerloo line extended from Elephant and Castle via Camberwell Green to Peckham. A new tube is included from Paddington via Marble Arch, Hyde Park Corner (or Lancaster Gate, Knightsbridge), Millbank and Waterloo to the Bank (replacing the Waterloo–City line).

B.20 Underground plan D comprises plan C with the new tube extended from the Bank via Liverpool Street, Whitechapel, Stepney East, Poplar, Canning Town, Custom House, Silvertown and Woolwich Central to Thamesmead. The Victoria line is extended from Brixton either to Streatham or Crystal Palace. The West End branch of the Northern line is separated from the City branch at Kennington and carried south to Stockwell and Clapham North. The Holborn–Waterloo line is extended to Vauxhall and Clapham Junction. The Leytonstone to Ongar and the Hainault parts of the Central line are handed over to British Rail.

B.21 Alternative bus systems were also studied. A possibility examined was the introduction of a network of high speed buses which might provide a better service than rail for certain journeys. These might run along the motorways, either on the normal traffic lanes or, if volume justified it, along special additional bus lanes.
In general, it was assumed that motorway buses would pick up and set down passengers at special stopping bays on the motorways, although no study was made of the physical practicability of this. On certain sections of route the buses were assumed to leave the motorways in order to serve important centres not directly on any motorway, such as Croydon. The radial routes did not terminate at Ringway 1 but were carried across the City and West End.
It was assumed that special or reserved routes might be provided for this purpose, but the practicability and cost of such provision were not examined. The speeds given to these motorway buses were calculated by taking the running speeds of private vehicles on the same roads and decreasing these speeds in proportion to the frequency of bus stops. As a result the speeds applied were unrealistically high.

B.22 It seemed logical to extend the current concept of Red Arrow services to provide a series of fast bus routes linking the main-line railway termini with Central Area destinations, and here again uncongested priorities were assumed to make possible speeds of 15 mph. These were referred to as 'standee' buses.
Other fast bus services were planned to link important transport interchange points on Ringway 1 with destinations in the Central Area, particularly those which were not already served by existing or proposed railway lines. The buses on these routes continued through the centre to the opposite side of the Ringway. One of the main functions of these buses was to serve park-and-ride traffic,

it being visualised that large-scale car parking would be provided at the transport interchange points.

B.23 A network of Green Line buses covered the whole of the LTS area serving areas untouched by the motorway buses. They only used motorways occasionally and operated in mainly radial directions passing through the Central Area. Their speeds were calculated in the same way as the motorway bus speeds.

B.24 All the bus networks examined contained the existing Red (Central) bus network of 1962 assumed to continue unchanged to 1981 except that several bad congestion points in the LTS area were assumed to have been removed.

Networks tested in Phase III

B.25 From these various public transport networks it was decided to choose British Rail plan C2 and Underground plan B for combining with a high speed bus system based on the plan 3 road network (motorways and base roads) to form a composite public transport network. This was called plan C2/B.

B.26 The various constituent parts of C2/B were designed with the assistance of the public transport Boards. The rail systems formed a practical and realistic service which could be operational in 1981. Conversely, the bus system was based on the assumption of free moving traffic in order to represent a highly efficient bus network. This might not be realistic but would enable the effect of such a system to be assessed.

B.27 A revised bus network with speeds compatible with the traffic volumes on the links was incorporated into public transport network C2/B/1, which while not being directly used in the test series, was used for some of the operational evaluation.

B.28 The alternative network used in the major comparisons in the test series was intended to improve the comfort and convenience of travellers into central London and was evolved by combining plan G (British Rail) with Underground plan B. This was called plan G/B.

C

Appendix C Costing of road networks

General

C.1 A considerable effort was made during Phase III to cost the alternative road plans and this was carried out by the consultants and various departments of the Greater London Council.

C.2 Estimates of the cost of property acquisition were provided by the Valuation and Estates Department of the GLC. In view of the magnitude of the task and the limited time available to complete it, some thought was given at the outset to the possibility of using methods such as applying values per mile based on typical lengths. It was found, however, that because variations between types and values of property were so wide, even over a short length of route, such a method of estimation could not be relied on. In the event, the Valuation and Estates Department adopted the same methods as those used when providing information on costs of road schemes for the Council and its Committees.

C.3 Estimates of cost of construction were prepared by the Construction Branch of the Department of Highways and Transportation of the GLC. The estimates were obtained from a library of current works costs, augmented by other contemporary published figures and information obtained from neighbouring authorities. In general, varying rates were used according to the type and location of each part of the works, depending on the width of the road and whether it was at ground level, depressed or elevated.

C.4 These estimates were based on alignments and designs prepared by either the authority responsible for the route in question or their consultants where design work had already started. In cases where designs were not already available, a preliminary route alignment was generally selected from Ordnance Survey plans and a field reconnaissance carried out. The route, including interchanges, was then generally drawn out on 1 : 2500 Ordnance Survey plans and examined by engineers and town planners. In this process development plans, geological maps and general data relating to existing structures were taken into account. The drawings were then completed showing the land and property affected by the route and the form of construction of the motorway links and interchanges.

Land acquisition and costs

C.5 The nominal acquisition width was primarily determined by the planning guidelines (see Appendix D), modified where necessary by the form of construction. The nominal width was increased, for example, by the necessity to acquire the complete residential lots affected.

C.6 The cost of land acquisition was made up as follows:

a. Purchase of privately owned property at current market values—e.g., houses, farms, private golf courses, statutory undertakers' property, abandoned railway tracks, private business premises.

b. Purchase of publicly owned property, the cost being that of replacement—e.g., parks, schools, hospitals, common land, municipal golf courses.

c. Compensation for 'injurious affection' of property—e.g., severance, and the dislocation of docks, wharves and river traffic during construction of immersed tube tunnels.

d. Easements for bored tunnels and viaducts with particular reference to railway tracks.

C.7 Notional costs for existing roads included in the acquisition band width were not included. All improvements to the existing motorways were assumed to take place within the existing reservations. The possible resale of lands becoming surplus to requirements, and the sale of air or other rights were not taken into account.

C.8 The variation in average land and property costs according to distance from central London is shown in Figure C.1.

Route design and construction costs

C.9 The route location of the road networks in plan 3 and in the selected routes within Ringway 1 has been discussed in Appendix B. The planning guidelines adopted are shown in Appendix D and construction details for plan 3 are shown in Table C.1. The total mileage is broken down by type of construction, number of lanes per carriageway and route location. Of the plan 3 motorways 48 per cent of the total miles were at ground level, 31 per cent elevated on viaduct or embankment and 19 per cent were depressed in retained or open cut. Tunnel construction accounted for the remaining 2 per cent.

C.10 The motorway link construction costs allowed for the following major factors:

a. *Width of carriageway*

b. *Factors varying with location*

services, drainage, river and road diversions, overbridges and underpasses, maintaining road and rail traffic during construction and support of existing structures during the construction of retaining walls.

Table C.1 Motorway construction details (miles)—plan 3

Type of construction	*No. of lanes*	*Ringway 1*	*Inner radials*	*Ringway 2*	*Outer radials*	*Ringway 3*	*Total miles*	*Percentage of total mileage*
Elevated (viaduct)	dual 2	—	—	—	0·25	—	0·25	
	dual 3	—	8·71	0·85	13·03	8·90	31·49	
	dual 4	18·30	13·37	14·77	8·65	1·79	56·88	
	Totals	18·30	22·08	15·62	21·93	10·69	88·62	25·5
Elevated (retained and embankment)	dual 2	—	—	—	0·31	—	0·31	
	dual 3	—	0·51	0·92	4·40	5·66	11·49	
	dual 4	0·43	0·45	3·31	4·46	—	8·65	
	Totals	0·43	0·96	4·23	9·17	5·66	20·45	5·9
Ground level	dual 2	—	—	—	2·80	—	2·80	
	dual 3	—	7·06	3·26	37·22	61·71	109·25	
	dual 4	3·24	6·92	12·79	24·93	6·98	54·86	
	Totals	3·24	13·98	16·05	64·95	68·69	166·91	48·1
Depressed (open)	dual 3	—	0·55	0·41	1·88	6·82	9·66	
	dual 4	1·63	—	0·10	6·50	0·97	9·20	
	Totals	1·63	0·55	0·51	8·38	7·79	18·86	5·4
Depressed (retained)	dual 3	—	4·66	1·31	6·83	6·76	19·56	
	dual 4	6·22	5·36	10·22	4·63	—	26·43	
	Totals	6·22	10·02	11·53	11·46	6·76	45·99	13·2
Tunnel (cut and cover) $<$1000 ft	dual 3	—	—	0·07	—	—	0·07	
	dual 4	0·13	0·10	0·36	—	—	0·59	
	Totals	0·13	0·10	0·43	—	—	0·66	
Tunnel (cut and cover) $>$1000 ft	dual 3	—	1·12	—	—	—	1·12	
	dual 4	0·63	0·38	0·90	—	—	1·91	
	Totals	0·63	1·50	0·90	—	—	3·03	1·9
Tunnel (bored and immersed)	dual 2	—	—	—	—	0·89	0·89	
	dual 3	—	—	0·99	—	—	0·99	
	dual 4	0·85	—	—	—	—	0·85	
	Totals	0·85	—	0·99	—	0·89	2·73	
Total mileage of links and interchanges		31·43	49·19	50·26	115·89	100·48	347·25	100

Furthermore variations occurred, such as the height of fill in embankments and the depth of excavation in cuttings, which affected the unit costs.

C.11 The average construction costs required for the motorway links in plan 3 are given in Figure C.2. It must be emphasised that unit costs were greatly influenced by particular circumstances. The situation arose where, for example, a section of dual-3 lane motorway with complex ancillary structures cost more than a dual-4 lane for the same nominal form of construction occurring elsewhere on the particular route location. The areas of land required for each form of construction were also affected by local conditions.

C.12 The average motorway costs per lane mile both of land acquisition and construction are tabulated in Table C.2, where the costs of motorway junctions with radials are included in the ringway costs. The average total cost of Ringway 3 was £230,000 per lane mile increasing to £1,350,000 per lane mile for Ringway 1.

C.13 The interchanges were motorway junctions, providing motorway to motorway connections and local access points connecting the motorway network to the local road system, or a combination of

Table C.2 Average motorway link costs* by location—plan 3

Location	*Land and property*	*Construction*	*Total*
Ringway 1	0·423	0·927	1·350
Inner radials	0·439	0·670	1·109
Ringway 2	0·205	0·498	0·703
Outer radials	0·084	0·260	0·344
Ringway 3	0·029	0·201	0·230

*Costs in £ million per lane mile (at 1966 prices)

Figure C.1 Average land and property acquisition costs by location—plan 3

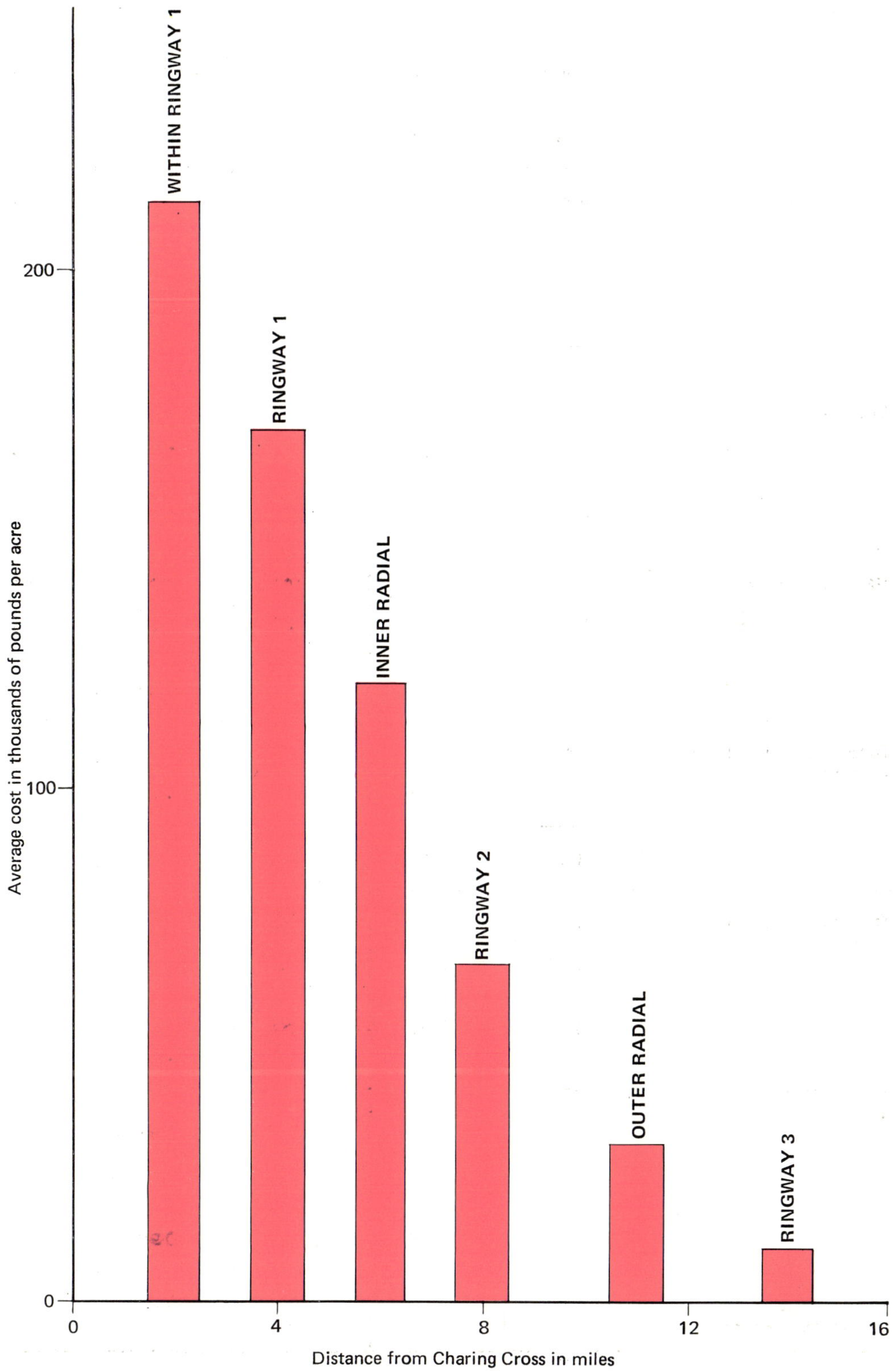

Figure C.2 Average construction costs per lane mile for all links—plan 3

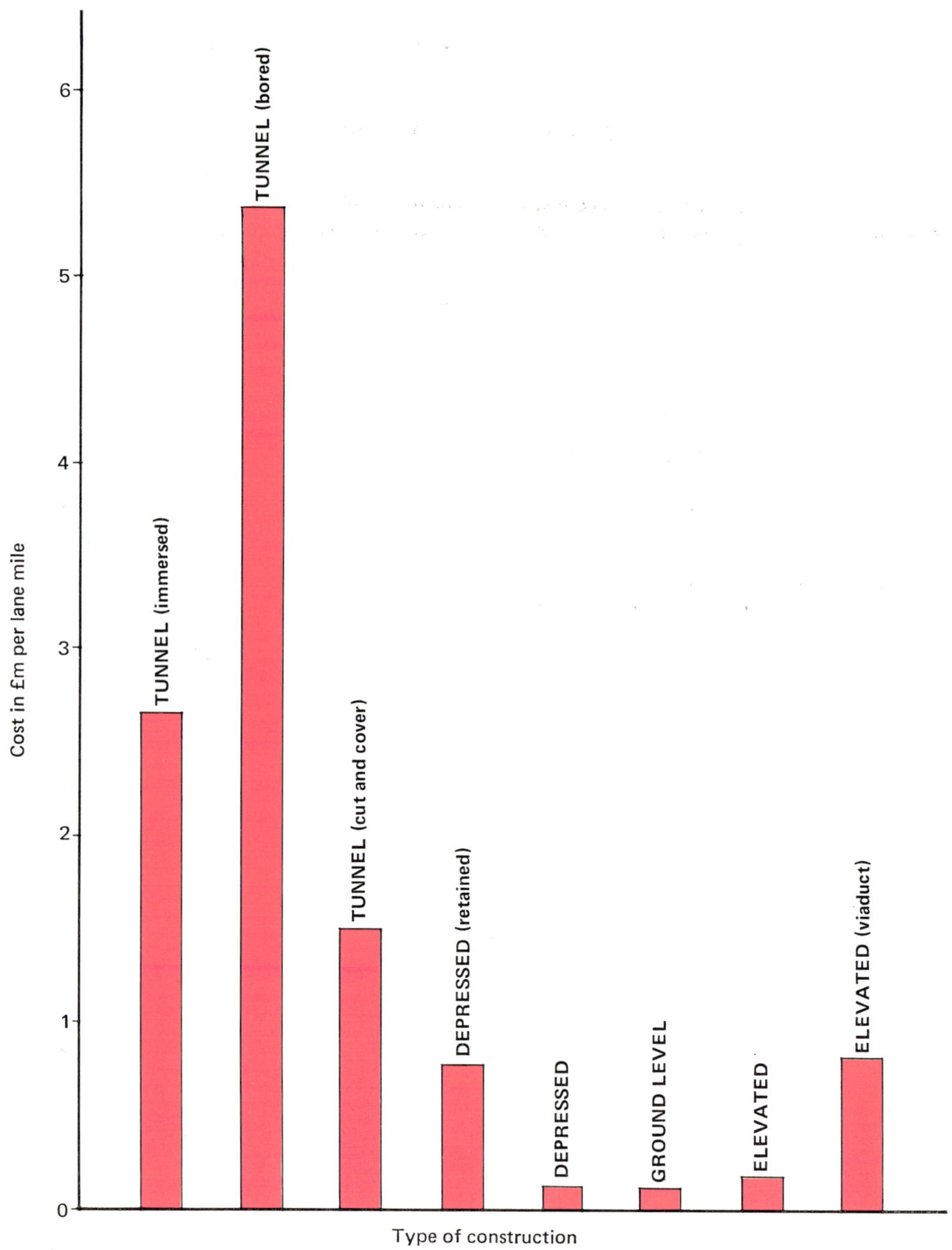

both. The average spacing, from centre to centre of adjacent interchanges, varied from 1.3 miles on Ringway 1 to 2.9 miles on Ringway 3.

C.14 The motorway junctions in particular required considerable areas of land. The average areas of the motorway junctions varied from 41 acres on Ringway 1 to 103 acres on Ringway 3. These ranges are illustrated in Figure 4.5.

C.15 The average land and property costs, and construction costs, for interchanges are given in Table C.3. In total they range for motorway junctions from about £5 million on Ringway 3 to £15 million on Ringway 1. The respective figures for local access points were £2 million on Ringway 3 and £9 million on Ringway 1.

C.16 The total land and construction cost for the 347 miles of plan 3 was estimated to be some £1,650 million of which 30 per cent was land and property costs and 70 per cent construction cost. Of the construction cost, about 69 per cent was due to interchanges.

C.17 It should be noted that these estimates of cost and land area relate to proposals which had uniformly generous environmental and geometric standards. It cannot be assumed that they represent actual costs of construction. Indeed the East Cross Route of Ringway 1 is now under construction at lower unit costs than were calculated in the LTS.

Table C.3 Average interchange costs*—plan 3

Location	*Land and property*		*Construction*		*Total*	
	Local access	*Motorway junctions*	*Local access*	*Motorway junctions*	*Local access*	*Motorway junctions*
Ringway 1†	4·5	6·75	4·5	8·75	9·0	15·50
Inner radials	2·0	2·75	2·75	3·75	4·75	6·50
Ringway 2†	1·25	3·50	2·50	8·0	3·75	11·50
Outer radials	0·75	0·25	2·75	3·0	3·50	3·25
Ringway 3†	0·25	0·50	1·5	4·5	1·75	5·0

*Costs in £ million (at 1966 prices)

†Costs of ringway—radial junctions included in the ringway costs

D

Appendix D Planning guidelines

D.1 Guidelines were used in Phase III in deciding motorway routes and forms of construction for the purpose of achieving uniform standards for costing alternative motorway networks.
These represented very high standards which may not necessarily be able to be achieved. Section A relates to form of construction and clearances, Section B to location.

Section A:

a. The appearance of the motorway and its ancillary structures must satisfy reasonable aesthetic standards.

b. Environmental considerations are likely to be the overriding factor in deciding both the alignment and the mode of construction of both the motorway itself and its points of access.

c. In deciding the mode of construction and the alignment the aim should be to produce the proposal which would represent the preferred long-term solution for that particular area. The cheapest solution to meet current conditions is unlikely to be adopted.

d. In residential areas the motorway should preferably be concealed, and normally depressed, thus minimising the nuisance of noise, headlights, visual intrusion and fumes.

e. The edge of the pavement should be at least 60 feet from the nearest dwelling if possible. If the near edge of the pavement is less than 60 feet away, then the motorway must be covered for 100 feet either side of the dwelling, in which case in no circumstances should the edge of the pavement be closer than 10 feet to any dwelling.

f. Where the motorway follows the line of an existing barrier to movement, e.g., an elevated railway line, then the motorway may be elevated although it passes through a residential area. However, every effort should be made to minimise adverse environmental effects and the clearances mentioned above should be maintained.

g. Where the motorway is depressed the clearance standards may be relaxed. However, the distance of the dwelling from the edge of the excavation or retaining wall of the depressed route should be greater than 40 feet.

h. In areas of public buildings or urban open space the motorway should preferably be concealed and normally depressed.

j. Owing to the necessity of providing for the cost of replacement of public open spaces it is not always economical to follow what may appear to be a 'soft' route.

k. In shopping areas and industrial sites any form of construction may be used, but the edge of the pavement must not be closer than 20 feet to any building.

l. In rural areas the motorway should normally be at ground level except when the edge of the pavement is within 100 feet of existing or proposed development.

m. The public generally prefer cemetery land to be avoided. It is difficult to acquire and should be avoided if possible.

n. Access to existing property should be given where possible to 'development control' standards and, where not possible, to a standard not less than that which existed before construction of the motorway.

Section B:

Motorway reservations should minimise the loss of or interference with:

a. Listed buildings (six hundred in former LCC area)

b. Ancient monuments

c. Structures and groups of structures of architectural, historical or other unique merit

d. Trees

e. Open space

f. Neighbourhood unity (where a major facility splits school, shopping, church and community-centre service areas)

g. Satisfactory pedestrian circulation patterns and safety due to increased traffic on access roads, etc.

h. Existing living or working standards due to noise, dirt, vibration

j. Visual amenity—disruption of neighbourhood unity by roads, street furniture (signs, etc.)

k. Housing

l. Property values due to noise, dirt, decline in upkeep, resulting in lowering of rates and thence loss of revenues available for community facilities and services

m. Cemeteries

n. Funds involved in revoking approved planning applications (not built) on the line of or near to the motorway route

o. Major sewers, piped streams and mains of statutory undertakers (gas, water, electricity, etc.)—particularly important where motorway is to be depressed

p. Farms or agricultural units where the severance of a strategic field or two could upset the balance of efficient working.

They should also maximise the benefit of:

a. Removal of substandard structures

b. Encouragement/opportunity of reinvestment in a decaying area by virtue of improved access facilities and elimination of substandard structures (including strips of land from road reserves being planted and landscaped) and of opportunities for profitable comprehensive redevelopment

c. Creating and rationalising neighbourhood units (school attendance areas, etc.)

d. Increase in property values.

E

Appendix E Major planning parameters

E.1 The major planning parameters used in the LTS and subsequently by the GLC (test 7) are summarised by sector in Table E.1. The 1962 totals are those reported in Volume I of the LTS and were used to calibrate the LTS model. The 1981 Phase II and III totals were reported in Volume II of the LTS and were the input which produced the results in Chapters 3 and 4 of this report. The 1981 GLC test 7 totals were the first revision produced by the GLC and were the input to the examination of the Council's Roads Objective network, described in Chapter 5.

E.2 Mention has been made in the text of the fact that the latest estimates for the Greater London Development Plan show further substantial reductions for 1981 of the population and employment totals in the GLC area (two-thirds the area of the LTS area). The implication of these new lower figures for traffic forecasts are in the process of examination.

Table E.1 Population (in households) and employment

	1962 (Phase I)		*1981 (Phases II, III)*		*1981 (GLC test 7)*	
Traffic sector	*Population*	*Employment*	*Population*	*Employment*	*Population*	*Employment*
0	1,348,000	639,000	1,333,000	744,000	1,338,000	664,000
1	941,000	368,000	976,000	432,000	968,000	358,000
2	221,000	101,000	263,000	115,000	266,000	125,000
3	1,229,000	460,000	1,187,000	511,000	1,201,000	491,000
4	1,370,000	662,000	1,393,000	656,000	1,253,000	639,000
5	191,000	1,339,000	199,000	1,516,000	200,000	1,381,000
6	30,000	105,000	31,000	120,000	31,000	124,000
7	1,432,000	504,000	1,562,000	518,000	1,502,000	522,000
8	581,000	169,000	618,000	188,000	646,000	207,000
9	1,162,000	425,000	1,253,000	517,000	1,230,000	502,000
Subtotal central area	221,000	1,444,000	230,000	1,636,000	231,000	1,505,000
Total LTS area	8,505,000	4,772,000	8,815,000	5,317,000	8,635,000	5,013,000

F

Appendix F Economic evaluation

Introduction

F.1 It is the aim of the economic evaluation procedure to compare the benefits to the community which would result from the improvement of transport facilities with the construction and other initial costs. In the LTS the comparison was made in the form of a one-year rate of return, that is the benefits in a single year, 1981, expressed as a percentage of the initial costs. The rate of return can be used to compare the relative merits of alternative plans.

F.2 The principal items giving rise to costs and benefits were:

Initial Costs

construction
land
other

Annual Costs and Benefits

users
operators
other

Initial costs

F.3 The procedure for costing the road plans examined is described in Appendix C. All costs were estimated at 1966 prices. All estimates of the cost of land and property required to facilitate construction were based on estimates made by the GLC's Valuation and Estates Department. Private land and property was valued at market value and publicly owned land and property at an estimated replacement cost. The only other initial costs specifically taken into account were those related to the provision of parking spaces for the requirements indicated by the plan tested.

F.4 Several factors were omitted from the analysis because of the difficulty of quantification. In particular noise, nuisance and traffic delays during construction and the costs of improvements to the existing road network were excluded.

F.5 Included in the land and property acquisition costs were estimates of the number of houses which would need to be demolished and the consequent displacement of people. It should be noted though, that the estimated costs include the costs of rehousing and it would be double counting to consider them as separate additional items.

Annual costs and benefits

Calculation of the benefits to transport users

F.6 The benefits to transport users were taken to include all those benefits initially received by transport users resulting from the completion of a scheme, irrespective of whether such users are the ultimate beneficiaries or not. For example some benefits might be taken up in reduced cost of goods and services resulting from improved accessibility. It would again be double counting to include these as well as the users' benefits.

F.7 The measure of this benefit in the LTS is essentially a measure of change in consumers' surplus. To estimate the benefit of a single plan and hence, by difference, the change in benefit between two plans, it is necessary to consider a demand curve for travel. This curve relates the demand q to given unit costs of travel c and is illustrated in Figure F.1.

F.8 The definition of the curve indicates that for any unit, say q′, up to the last unit of q, the buyers would have been willing to pay c′, thus making a profit or surplus of c′—c. When all these surpluses are added they form the coloured area in Figure F.1 which is the consumers' surplus.

F.9 In fact when the cost to the users is apparently c, some part of this, t, is not a resource but is in fact taxation payment. Since this is merely a transfer, the total surplus is the coloured area shown in Figure F.2.

F.10 Where restraint on travel is imposed, of the form simulated in the LTS model by the linear programming procedure, then the system operates at a point R. This is shown in Figure F.3. The consumers' surplus is again given by the shaded area. The restrained situation could be brought about by direct or fiscal control but in either case, for the purposes of the LTS, the costs of administering the control were ignored.

F.11 When two plans are compared the net benefit is given by the difference in consumers' surplus and is illustrated in Figure F.4. The total change illustrated has been divided into three areas:

a. Benefit due to change in trip costs.

b. Benefit due to increased capacity of the system.

c. Benefit measured in terms of the tax paid by generated traffic.

F.12 The areas were calculated separately, with costs in time units. For the evaluation it was necessary to convert these to money units by the inclusion of personal time valuation and direct transportation or operating costs.

Personal time valuation

F.13 Time used or saved on business purposes was valued at an estimated average wage of 15/3d per hour for London in 1981, at 1966 prices, based

Figure F.1 Demand curve and consumers' surplus

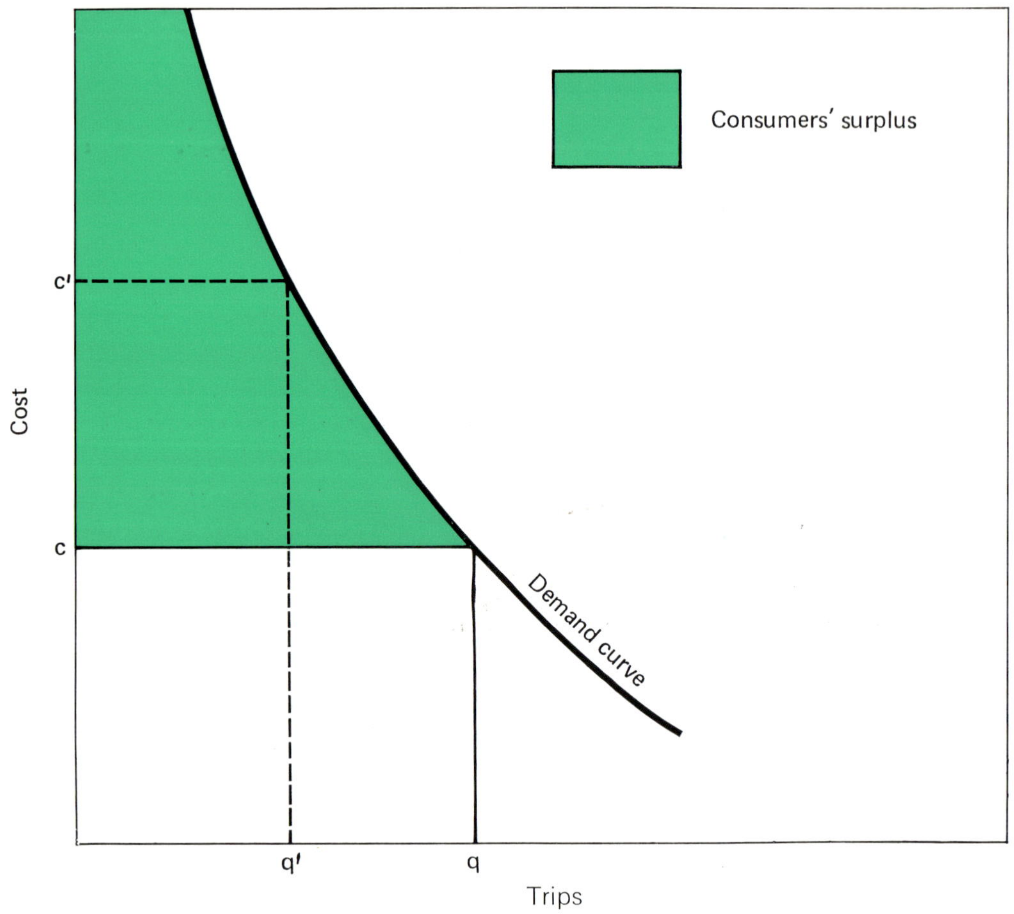

Figure F.2 Total surplus

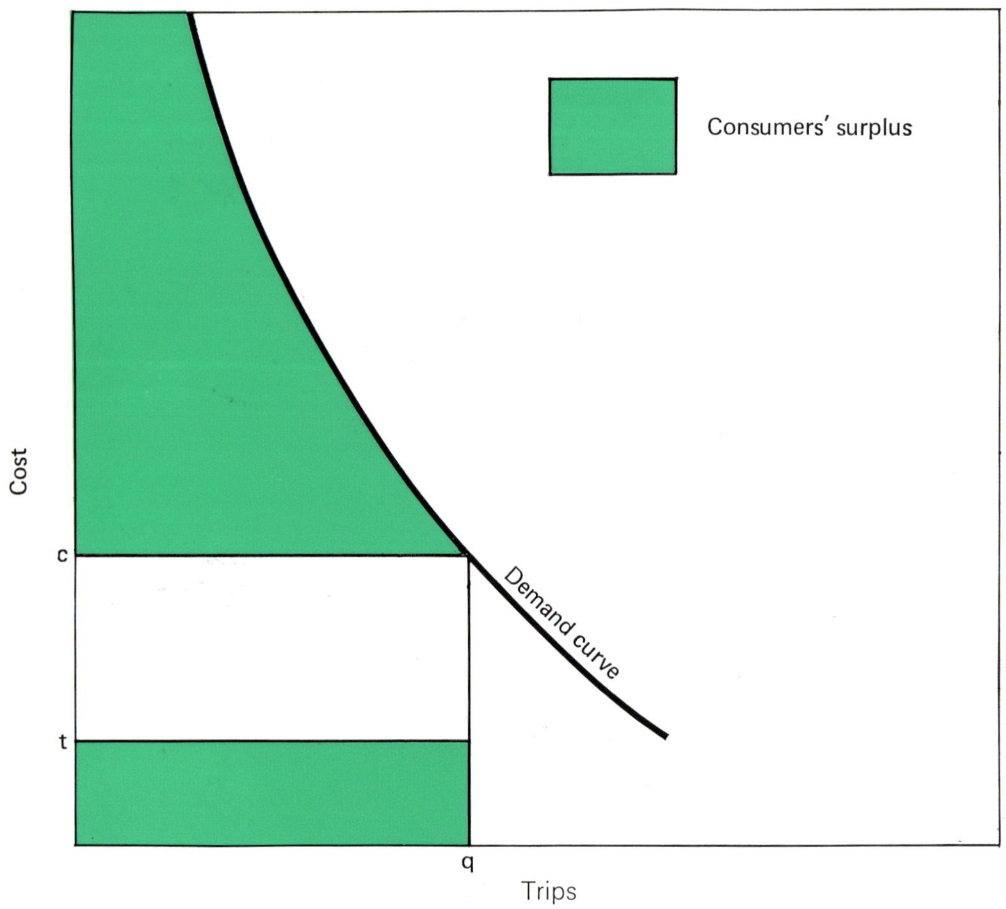

on an assumed increased in real income per head of 2·8 per cent per annum from 1962. All other travel time was accorded one-third of the value used for business purposes. The 1981 and corresponding 1966 figures are illustrated in Table F.1. By taking account of the proportion of trips of each type which make up each trip purpose, the personal time cost per person trip by purpose was calculated. The values are shown in Table F.2. Similar values per car vehicle trip were obtained by taking account of car occupancy.

Direct or operating costs

F.14 The Road Research Laboratory's Technical Paper No 75 'The Economic Assessment of Road Improvement Schemes' gives various unit costs relating to the cost of operating private cars, broken down into costs per mile and per hour. The depreciation element was excluded in the LTS on the grounds that depreciation of private cars is unlikely to be a function of the amount of travel. Indirect taxes, such as fuel taxes, were included and the resulting direct costs per car were 3·6d per mile and 0·72d per minute as shown in Table F.2.

F.15 A similar analysis to that for private cars was undertaken for commercial vehicles although in this case depreciation was included on a mileage basis and a specific London road haulage wage rate, including a 20 per cent allowance for overheads, was used. Table F.3 gives the resulting 1981 costs per mile and per minute used for commercial vehicles.

Table F.1 Value* of personal time savings by trip type

Trip type	*1966*	*1981*
Business	9/11d	15/3d
Commuting	3/4d	5/1d
All other	3/4d	5/1d

*Hourly value (at 1966 prices)

F.16 The fares paid for public transport vary considerably according to the category of user, service used, distance travelled, the time of day and regularity of travel. The range of fares per mile in 1966 was very wide, extending from fares of 4d per mile or part of a mile for short journeys to well under 2d per mile for regular long-distance commuters, to say nothing of the low cost or free travel offered to large numbers of public transport staff and their dependants.

F.17 On the basis of figures supplied by British Rail and London Transport the average fare in 1966 for all work trips by public transport was taken to be 2·72d per mile. This is considerably lower than the rate for non-work journeys which was taken to be 3·38d per mile. These were used unchanged for 1981 and are included in Table F.2.

Costs and benefits to operators

F.18 Costs of operating private and commercial vehicles were included under user costs and need no further discussion. Specific estimates of road maintenance costs were made which were, however, not functions of the amount of travel but simply of the network under test.

F.19 When the road plans were compared it was assumed that any decrease in public transport revenue would be compensated by reduced operating costs and/or increase in the comfort of the remaining customers using public transport. Neither were therefore specifically taken into account. In the comparison of the public transport plans estimates were made of the change in working expenses due to the rail lines, particularly the new BR deep level tunnels, and of bus operating costs.

Other costs and benefits

F.20 Many of the costs and benefits were estimated very approximately but were nevertheless taken into account. Two which were not included were changes in the quality of the environment and changes in accident costs.

Figure F.3 Restrained demand

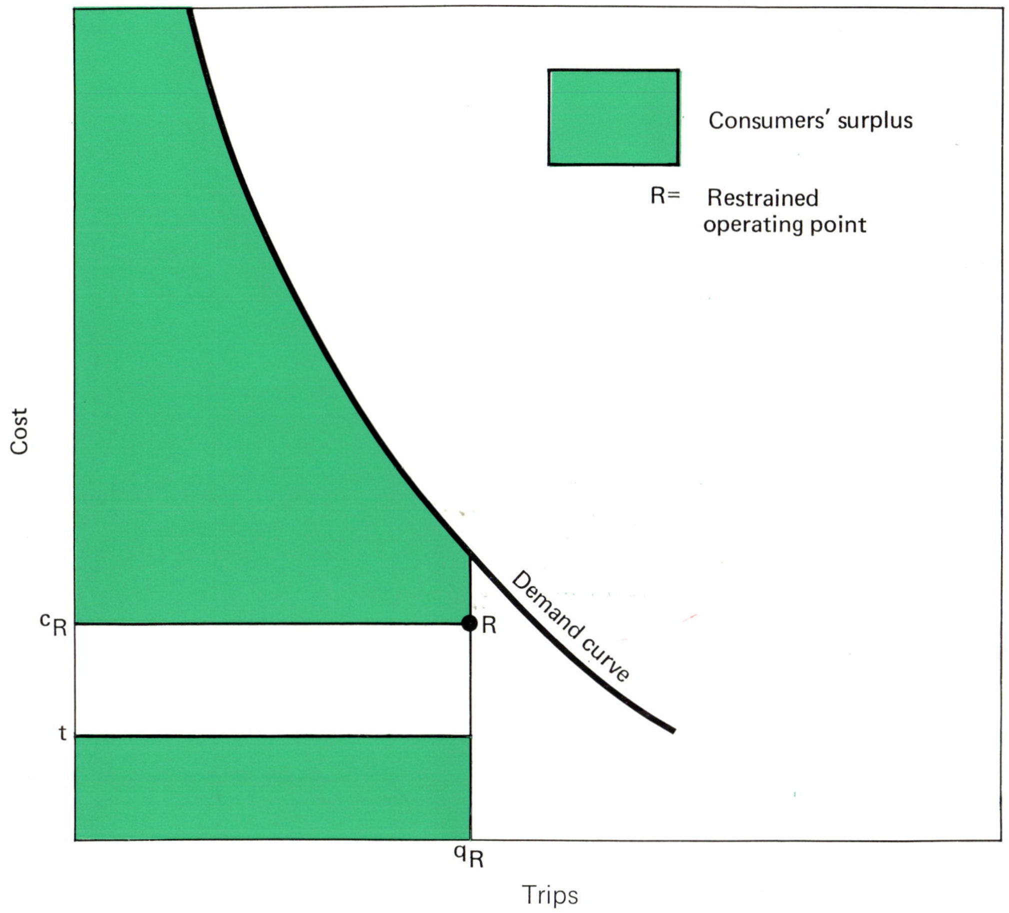

Figure F.4 Comparison of plans

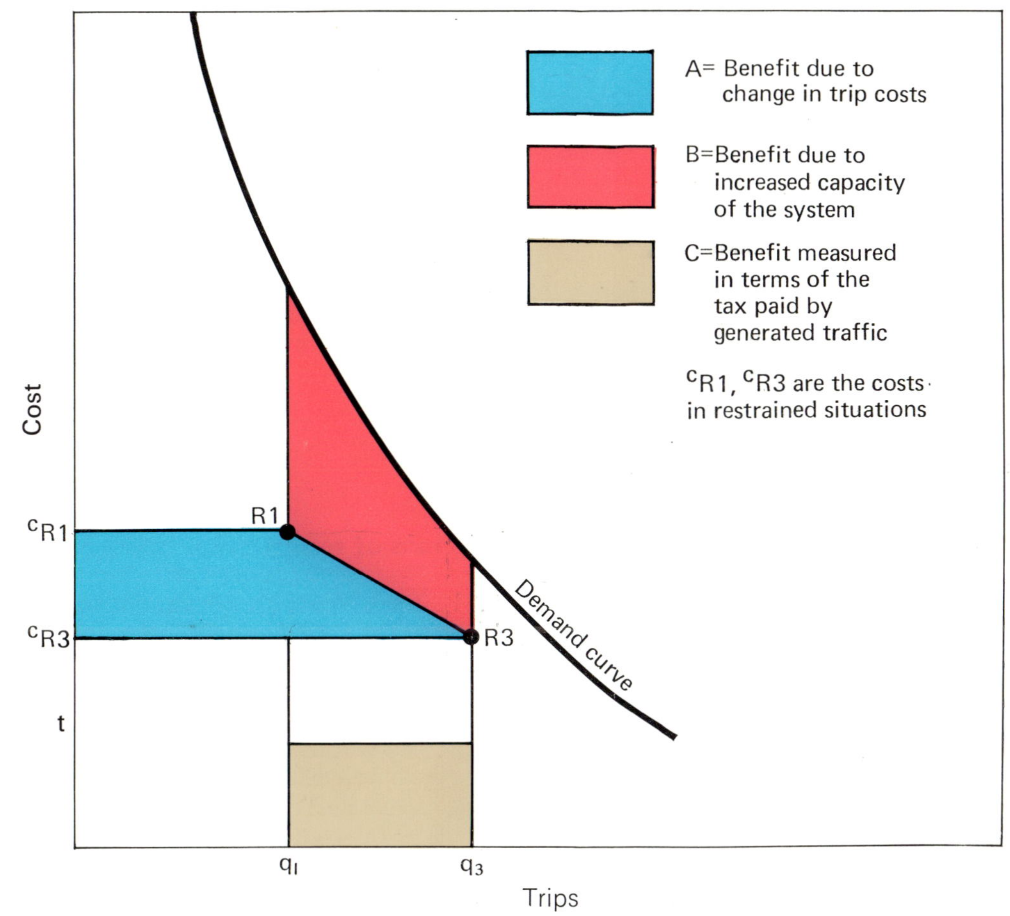

Table F.2 Total cost* per trip, by trip type and purpose in 1981

Category of trip cost	Car					Public Transport	
	Occupancy	per mile	Per minute			Per mile	Per minute
			Direct	Personal time	Total per minute		Personal time
Per person by trip type							
Business	1·13	3·19d	0·64d	3·05d	3·69d	3·38d	3·05d
Commuting	1·16	3·10d	0·62d	1·02d	1·64d	2·72d	1·02d
All other	1·70	2·12d	0·42d	1·02d	1·44d	3·38d	1·02d
Per person by trip type							
Work	1·16	3·10d	0·62d	1·02d	1·64d	2·72d	1·02d
Other home-based	1·67	2·17d	0·43d	1·12d	1·55d	3·38d	1·12d
Non-home-based	1·32	2·73d	0·54d	2·19d	2·73d	3·38d	2·19d
Per vehicle by trip type							
Work	1·16	3·60d	0·72d	1·18d	1·90d		
Other home-based	1·67	3·60d	0·72d	1·86d	2·57d		
Non-home-based	1·32	3·60d	0·72d	2·89d	3·61d		

*At 1966 prices

Table F.3 Operating costs* for commercial vehicles in 1981

Per mile	Per minute		
	Vehicle cost	Occupants' cost	Total
5·93d	0·59d	4·48d	5·07d

*At 1966 prices

It was decided that traffic should be assumed to be restrained to the level required to keep speeds on the base roads similar to those in 1962. In practice a certain amount of extra investment might be required in selected areas to maintain these speeds, for example adjacent to the motorways.

The rate of return

F.21 The LTS used a simple comparison between the annual benefit that would be obtained in a specific year, 1981, given the existence of a specific composite network, and the cost of building that network. This was because travel demand data was not available for any other future year. The interest on the capital involved in building the network and the time stream of benefits during the years up to and beyond 1981 were not calculated.

F.22 It should be noted that all the benefits and costs considered were taken at constant 1966 prices so as to avoid the difficulties of, and distortions due to, predicting future rates of inflation. This means that any costs which have and will continue to have a heavy proportion of labour costs, whether direct or indirect, have been underestimated, while the costs containing a low element of labour cost have been overestimated. It is also quite probable that land costs will increase at a rate different from the average of all other costs.

F.23 The one-year rate of return in 1981 at 1966 prices, was in all comparisons calculated as:

$$\frac{\text{Net benefit difference between plans in 1981}}{\text{Total cost difference between plans}}$$

G

Appendix G Trip pattern in 1981

G.1 The overall pattern of movement for the LTS area is subdivided in Figure G.1. The trips are divided into five types:

1. Internal trips with both ends in the Central Area.

2. Internal trips with one end in the Central Area.

3. Internal trips with both ends in the non-Central Area.

4. External trips with one end in the Central Area.

5. External trips with one end in the non-Central Area.

G.2 Trips have been divided into car and motor-cycle driver, commercial vehicle, taxi and public transport (reported and unreported) for the trips

Table G.1 Trip* pattern in Phase III test 4 (post-restraint) GLC test 7 (post-restraint)

	Private and commercial driver				*Public transport*		
	Car and motorcycle driver	*Commercial vehicle*	*Taxi*	*Total*	*Reported*	*Unreported*	*Total*
1.	0·16	0·15	0·10	0·41	0·15	0·21	0·36
	0·12	0·15	0·10	0·37	0·14	0·21	0·35
2.	0·64	0·22	0·04	0·90	2·07	0·26	2·33
	0·52	0·22	0·04	0·78	1·99	0·26	2·25
3.	6·53	1·39	0·04	7.95	3·64	0·35	3·99
	4·89	1·39	0·04	6·32	3·93	0·35	4·28
Subtotal	7·32	1·76	0·18	9·26	5·86	0·82	6·68
internal	5·53	1·76	0·18	7·47	6·06	0·82	6·88
4.	0·08	0·02	—	0·09	0·61	0·06	0·67
	0·08	0·02	—	0·09	0·61	0·06	0·67
5.	0·84	0·28	—	1·12	0·12	0·06	0·18
	0·84	0·28	—	1·12	0·12	0·06	0·18
Subtotal	0·92	0·30	—	1·22	0·73	0·12	0·85
external	0·92	0·30	—	1·22	0·73	0·12	0·85
Total	8·24	2·06	0·18	10·48	6·59	0·94	7·53
	6·45	2·06	0·18	8·69	6·79	0·94	7·73

*24-hour average weekday millions of trips. Car and motorcycle passenger trips excluded. External bus trips excluded; upper figures—Phase III test 4; lower figures—GLC test 7.

1. Internal trips with both ends in the Central Area.
2. Internal trips with one end in the Central Area.
3. Internal trips with both ends in the non-Central Area.
4. External trips with one end in the Central Area.
5. External trips with one end in the non-Central Area.

in Phase III test 4 (see Chapter 4) and Greater London Council test 7 (see Chapter 5) for each of the five movements. They are shown in Table G.1 where the upper figures refer to Phase III test 4 and the lower figures to GLC test 7.

G.3 It will be seen that the same matrix was used in both tests for internal commercial vehicle, taxi and unreported public transport trips and all external trips as has been indicated in Tables 4.10 and 5.12. Tables G.1, 4.10 and 5.12 all exclude car and motor-cycle passenger trips as well as external bus passenger trips whereas Table 2.10 for 1962 includes these trips.

G.4 The substantial decrease in driver trips and slight increase in public transport trips between Phase III test 4 and GLC test 7 was noted in Chapter 5. It is clear that the reduced primary road network in GLC test 7 considerably restrains movement on the roads. It is important to note however the similarity in the general pattern of movement in both tests, namely:

a. The small proportion of external trips.

b. The predominance of the car for trips outside of the Central Area; this is even more pronounced when car passenger trips are added.

c. The predominance of public transport for trips to the Central Area; this is even more pronounced for work trips alone.

G.5 The work in Phase III of the LTS and subsequent work by the GLC was designed to obtain more realistic estimates of the future trip pattern whereby the traffic flow was in balance with the capacity of the networks provided. Chapters 4 and 5 have described the implications of this work. None of the tests have suggested however that the nature of the transport problem in London so far as the overall pattern of movement is concerned is likely to be substantially different from that foreseen in Phase II of the LTS.

Figure G.1 Trip pattern

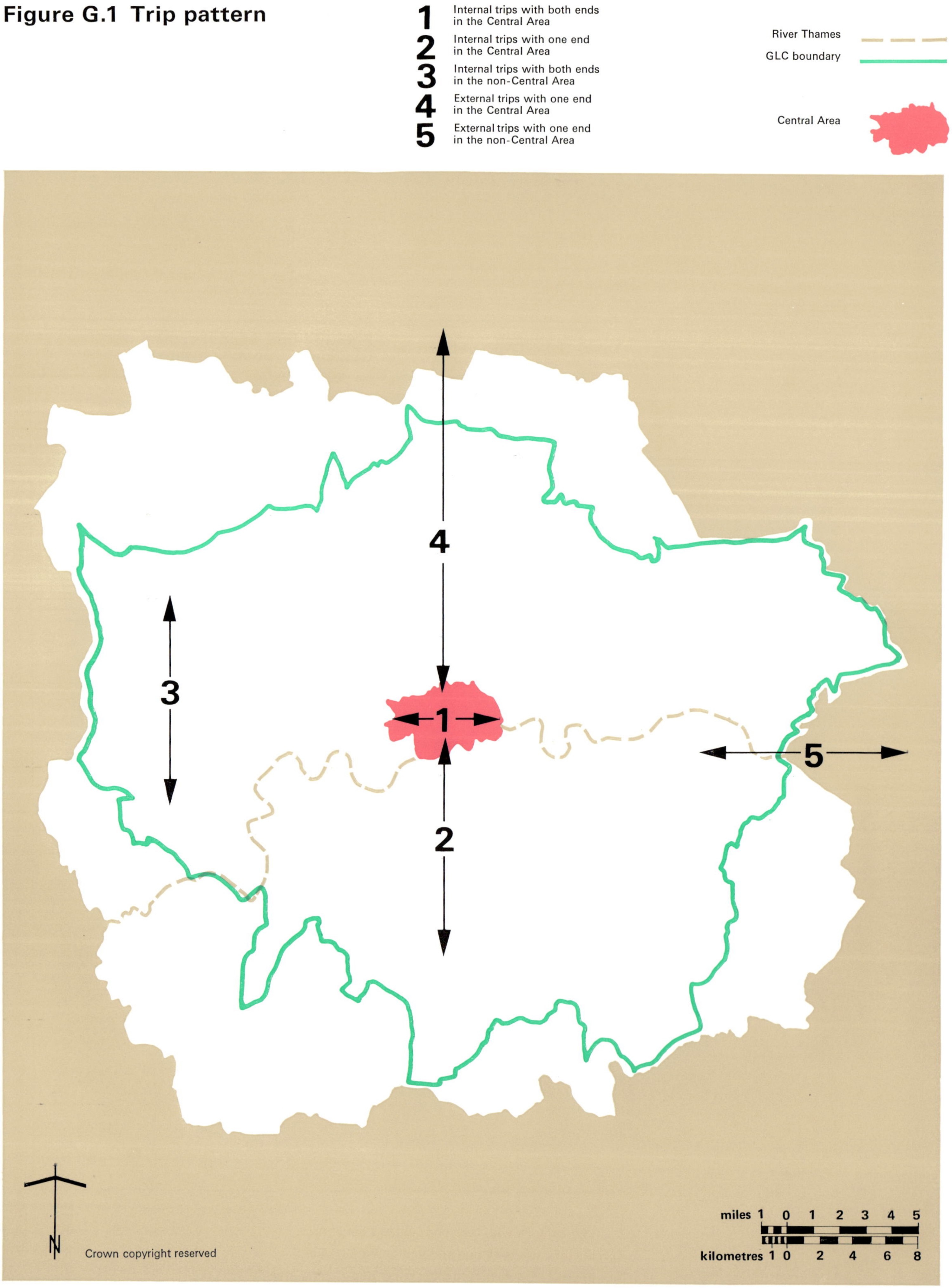

Glossary

Glossary of technical terms

Area capacity
A measure of the capacity of the secondary or base road system in an area. Estimated for jurisdictions, in terms of vehicle-miles per day.

Assignment
See 'Traffic Assignment'.

Attraction
The non-home end of a home-based trip; the destination of a non-home-based trip; the 'purpose' end of a trip.

Base road, secondary road
A road not classified as a motorway.

Basic trip
A one-way movement from one place to another for a particular trip purpose using one or more motorised modes. In modal distribution studies basic trips are classified by 'main mode of travel'.

Boarding penalty
A notional time which may be added for each change from one public transport service to another which seeks to simulate the disincentive associated with these changes.

Bus
London Transport central road services, country buses and other scheduled bus services.

Car driver
The driver of a car or motor-cycle.

Car occupancy
The average number of persons in a car, including the driver.

Car-owning household
A household in which at least one member has available for regular personal use a car in operating condition, whether owned outright, subject to hire purchase, borrowed, assigned by a company or public authority, or other.

Central Area
The same as the Greater London Conurbation Centre delineated as Division 1 for the 1961 Census; it contains an area of about 10 square miles, and is divided by the River Thames into two traffic sectors, viz. 'Central Area North' and 'Central Area South'; LTS sectors 5 and 6.

Centroid
A point which represents a traffic zone or district for traffic analysis purposes.

Centroid connector
A line along which trips from a zone or district are fed to a node on the network from a centroid.

Coach
School, works and chartered coaches, airline coaches.

Commercial vehicle
A mechanically propelled vehicle constructed or adapted for the use of and used for the conveyance of goods or burden in the course of trade or business, and licensed as such.

Cordon
An imaginary line which completely encloses a given area and at which traffic counts and interviews may be taken for control purposes.

Corridor
A group of related roads or public transport services serving similarly oriented traffic movements.

Cycle, motor-cycle
Any motor propelled cycle, moped, scooter, motor-cycle or combination.

Design volume
The volume used to select the number of lanes desirable on a motorway in order that operating conditions do not deteriorate to below a chosen level. See 'Limiting volume'.

Distribution
See 'Trip distribution'.

District
See 'Traffic district'.

Diversion curves
Curves showing the proportion of trips made by public transport for different ratios of the time taken for a trip by public transport to the time taken by private transport, between given origins and destinations. Used to estimate choice of mode.

Excess time
The time spent waiting and walking while making a public transport trip.

External cordon
The cordon delineating the boundary of the LTS area.

External trip
A trip with one end inside and the other outside the LTS area.

Flow
See 'Volume'.

Generation
The home end of a home-based trip; the origin of a non-home-based trip.

GLC
The Greater London Council.

GLDP
Greater London Development Plan.

GNP
Gross National Product.

Heavy commercial vehicle
A commercial vehicle not classified as a 'Light commercial vehicle'.

High-speed bus
A bus operating at higher speeds than are currently normal which may require special road facilities.

Household
A person living alone or a group of persons living together, partaking of meals prepared together and benefiting from common housekeeping.

Income
Total gross income, before tax or other deduction, of all household members received from all sources, including wages or salary for employment, profits and salary from business, pensions, Social Security, sickness and unemployment benefits, bonuses, dividends and rent; expressed as an annual value.

Interchange (motorway)
A location on a motorway where it is possible to change from one route to another. (For types of interchange, see 'Motorway junction' and 'Local access point').

Interchange (public transport)
A point at which it is possible to change from one public transport service to another.

Internal trip
A trip with both origin and destination inside the LTS area.

Inter-zonal trip
A one-way movement between two zones.

Intra-zonal trip
A one-way movement within a zone.

Jurisdiction
One of four areas into which the LTS area was divided. Used primarily in conjunction with area capacities. The areas are:
- Central Area
- Remainder of the former LCC area
- Remainder of LTS area north of the Thames
- Remainder of LTS area south of the Thames.

LCC
The former London County Council.

Light commercial vehicle
Generally a commercial vehicle having a van type body and only two rear wheels; usually a vehicle of not over 30 cwt unladen weight.

Limiting volume
The maximum volume which could be accommodated on a facility without undue risk of breakdown in flow at peak periods; it is somewhat less than the absolute maximum capacity of a facility.

Linear program(me)
A procedure by which a function is either maximised or minimised subject to defined constraints. Both the function and the constraints are linear. In the context of this report linear programming was applied so as to maximise the number of private transport trips, the most important constraint being the capacity of the road system.

Link
A representation of a facility connecting two nodes in a network.

Loading
That part of the traffic assignment process in which the zone to zone or district to district trips which form the trip matrix are allocated to the corresponding trees of the network and in which, principally, the traffic volumes assigned to each link in the network are accumulated.

Local access point
An interchange between a motorway and a secondary road.

LTS
The London Traffic Survey or the London Transportation Study.

LTS area
The London Traffic Survey Area: a 941 square-mile area containing all of the Greater London Council area and portions of the counties of Essex, Hertfordshire, Kent and Surrey; the area to which most of the various studies described in this report relate.

Matrix
See 'Trip matrix'.

Minimum path
The route of travel between a pair of traffic zones or districts which has the least accumulation of time, distance or other parameter. Minimum time is the parameter used throughout this study.

Modal choice
The proportions of trips using various modes of travel.

Mode of travel
A form of transport used for a person trip. In the LTS the term generally refers exclusively to one of the following motorised modes: car driver, car passenger, cycle driver, cycle passenger, bus passenger, Underground passenger, British Rail passenger, taxi passenger and commercial vehicle driver. Trips on foot and by pedal cycle are generally excluded from studies of modal distribution, in this report.

Motor-cycle
See 'Cycle'.

Motorised mode
See 'Mode of travel'.

Motorway
A dual carriageway road of high design standards with grade-separated interchanges. It does not necessarily comply with the statutory requirements for a Special Road, however. Synonymous with 'Primary road'.

Motorway junction
An interchange between two or more motorways.

Network
A road, rail or other transportation system; an arrangement of links and nodes which represents such a system.

Node
A numbered point which defines the end of a link in a network. In a road network each node generally represents either a road intersection or a centroid and in the public transport network a bus stop or railway station or, in either case, the point at which a centroid connector joins the network.

Non-car-owning household
A household, no member of which has regular use of a car, as defined under 'Car-owning household'.

Non-Central Area
All sectors of the LTS area other than those in the Central Area.

Non-resident
A person not living in or based in the LTS area.

Off-peak hours
The hours other than those in the morning and evening peak periods.

Parameter
See 'Planning parameters'.

Pavement
The structural part of the road including the central reserve, the carriageways and any hard shoulders, kerbs or parapets.

Peak hour
An hour, usually an integral hour, of maximum traffic flow.

Planning parameters
Statistics concerning population, employment, car ownership, business and retail centres, schools and school attendance, land use densities and other data.

Potential car driver
A person travelling from a car-owning household, or from a non-car-owning household with access to a car or motorcycle. (Car and motorcycle passengers excluded.)

Primary road
See 'Motorway'.

Private transport (trips)
Travel by private vehicles. Excludes travel by taxis or commercial vehicles.

Program
A sequence of coded instructions, usually stored on magnetic tape, used to solve a problem by electronic computer.

Public transport
Mass transport along fixed routes; applies to buses, Underground and British Rail.

Public transport (trips)
Passenger travel by public transport vehicles.

Purpose
See 'Trip purpose'.

Regression analysis
A statistical technique which, for an assumed mathematical relationship between two or more variables, determines the coefficients which produce the best fit with observed measures of the variables.

Resident
A person living in or based in the LTS area.

Route (public transport)
A road or railway along which one or more public transport services operate.

Screen line
An imaginary line, generally following a natural barrier, at which traffic counts or interviews may be taken for control purposes.

Secondary road
See 'Base road'.

Sector
See 'Traffic sector'.

Service (public transport)
A particular public transport service operating between specific points in the network (e.g., No. 11 bus).

Standee bus
A bus designed for a majority of standing passengers (e.g., the London Transport 'Red Arrow' bus).

Taxi, taxi cab
A vehicle licensed as a hackney carriage, with four seats excluding the driver.

Traffic assignment, assignment
The process whereby trips from an origin to a destination are assigned (allocated) to definite routes or public transport services (existing or planned), based on factors known to influence route selection, the primary factor being travel time.

Traffic district, district
A group of traffic zones, designed to facilitate analysis of the Survey data; there are 186 internal traffic districts, averaging about five zones per district in the LTS area.

Traffic Model
A series of mathematical formulae expressing the actions and interactions of travel and land use and other planning parameters in such a manner that travel patterns can be synthesised or estimated from given assumptions about these factors.

Traffic sector, sector
A group of traffic districts. The LTS area is divided into ten internal sectors.

Traffic zone, zone
The basic unit of area for traffic analysis, delineated with regard to the main road system, major barriers to traffic flow, land use characteristics and other traffic engineering considerations; there are 933 internal traffic zones in the LTS area.

Travel time
The average total running time between one node and another, including delays, in a network. A typical road travel time between one centroid and another consists of an initial time spent on a centroid connector, an average running time on roads included in the road network, a terminal time on a centroid connector and, in certain circumstances, a notional parking time. A typical public transport travel time consists of an initial time on a centroid connector, waiting times, travel times on services, walking times between services, possibly boarding penalties and terminal time on a centroid connector.

Tree
A record of the minimum paths from one centroid to all other nodes and centroids in a network.

Trip
A one-way movement from one place to another for a particular purpose. As used in this report, the term generally refers to movements using motorised modes.

Trip distribution, distribution
The mathematical procedure by which the trip matrix is created from a knowledge of generations and attractions in each zone or district and the travel time for each zone to zone or district to district movement.

Trip end
The origin or destination of a trip.

Trip matrix, matrix
A record of zone to zone, district to district or sector to sector trips, giving the numbers of trips in each direction between each pair of zones, districts or sectors.

Trip purpose
The motivating purpose for which a trip is undertaken.

Underground
London Transport rail services, whether above or below ground level.

Unlinked trip
A one-way movement from one place to another for a particular trip purpose by a particular mode of travel, as recorded on the home interview form, and which is considered to be completed when either the trip purpose has been accomplished or a change in mode of travel has occurred.

Unreported trips (public transport)
The public transport trips not reported in the London Traffic Survey.

Vehicle
Vehicles travelling on roads including cars, cycles, taxis, and commercial vehicles.

Volume, flow
The number of vehicles passing a given point in a specified period of time.

Waiting time function (public transport)
The function which gives the average time spent by the prospective passenger waiting for a public transport service (excluding walking time).

Zone
See 'Traffic zone'.